DISCOVERING SCIENCE

HOUGHTON MIFFLIN COMPANY
BOSTON
DALLAS GENEVA, ILLINOIS HOPEWELL, NEW JERSEY
PALO ALTO LONDON

DISCOVERING SCIENCE

JOHN W. HARRINGTON
WOFFORD COLLEGE

Cover photograph by James Scherer

Selected material in Chapter 2 is quoted from Thomas S. Kuhn's *The Structure of Scientific Revolutions,* 2nd ed., International Encyclopedia of Unified Science (Chicago: University of Chicago Press, 1970), II. Copyright © 1962 by the University of Chicago Press. All rights reserved.

Material on pages 114 and 115 is reprinted from *Exchange and Production: Competition, Coordination, and Control,* Second Edition, by Armen Alchian and William R. Allen. © 1977 Wadsworth Publishing Company, Inc., Belmont, California 94002. Reprinted by permission of the publisher.

Portions of the present volume incorporate material from the author's book *To See a World* (St. Louis: C. V. Mosby Co., 1973).

Printed in the U.S.A.

Library of Congress Catalog Card Number: 80-80721

ISBN: 0-395-25527-9

TO MARTHA WHARTON,

INDEFATIGABLE REFERENCE LIBRARIAN,

WHO FOUND THE FACTS,

AND DAN W. OLDS,

PHYSICIST, WHO KNEW HOW TO USE THEM.

CONTENTS

BEFORE WE BEGIN

The mission of this book is to make its readers better students by showing them how scientists think, how they discover the system and order in nature, how they use units and measurements, and how they become motivated in their professions. These are exactly the same skills that students and teachers must acquire to be able to do their own work. *Discovering Science* offers the opportunity to learn these things without assuming, as so many scientific books do, that the reader already knows them.

Every discipline in the physical, life, and social sciences shares a common purpose and fabric with all of the others. Each one is dedicated to the same task, *the progressive discovery of the nature of nature.* Unfortunately, beginning science students and their teachers are usually held within the bounds of a single discipline and forced to deal with a vast amount of detail in a very short period of time. This tends to rob them of the opportunity to see or show the unity of science as a model of the unity of nature. *Discovering Science* is designed to overcome that drawback in our educational system by weaving together classic illustrations from many different disciplines.

This technique helps make the book applicable to the full spectrum of science and social science courses. *Discovering Science* is intended to be used as a supplement to beginning courses in physics, astronomy, chemistry, geology, biology, psychology, geography, economics, sociology, anthropology, archaeology, and government. It can also be used as a supplement to the suite of education courses that are involved with the problems of teaching methods in all of these areas.

The test of the book is a simple one. Does reading it offer a better understanding of the scope of all science and at the same time increase the desire to master the details of a specific area? I hope that instructors will put the book to this test with their students.

A brief preview of the text will spell out how an understanding of science is developed. The principal strategy is to offer familiar points of reference to serve as benchmarks and make both teachers and students feel at home when they move from one science into all sciences simultaneously.

Chapters 1 and 2 explain the game of science and how it is won. Illustrations from the fields of astronomy, biology, and geology are cited in the work of Galileo, Leonardo da Vinci, Charles Darwin, James Hutton, and Alfred Wegener.

Chapter 3 deals with the problems of classification and the search for system and order in nature. The major illustrations are taken from the history of science. Each one of them was a revolution in its day. Carolus Linnaeus is shown grappling with the problem of classifying a bewildering number of different plants and animals. Dmitri Mendeleev and Lothar Meyer are seen struggling to grasp the pattern of the periodic table of chemical elements. Knowledge of system and order is one measure of scientific maturity at all levels. Students are often most aware of their own inadequacies when they find themselves fumbling with an incomplete appreciation of the system and order that underlies scientific nomenclature. At the professional level, the inability to fit facts into an inclusive context demonstrates scientific immaturity. The great breakthrough of plate tectonics that swept geology in the early 1960s eliminated so many questions and raised the science to a level of maturity it had never reached before. Comparison with the development of psychology and the social sciences indicates that these disciplines are in the very early stages of establishing an understanding of the system and order in the parts of nature with which they deal.

Chapter 4 is about the role of quantitative measurement and the use of precise units. Quantitative thinking begins with the act of counting things. An illustration from the field of economic geography shows the kinds of insights that may be based on very simple data. The beauty of quantitative reasoning is displayed in another geographic illustration by following the reasoning of Eratosthenes, who actually measured the size of the spherical earth in the third century B.C. More complex kinds of units of measurement are examined in several social sciences.

The last part of Chapter 4 discusses the idea of learning to view nature in proper scale. One of the most important aspects of this skill is that of answering complex questions through the method of making successive approximations. Sample calculations show how to estimate the number of grains of sand on a beach and the number of fish in all the oceans of the world. A final illustration of gaining understanding by viewing things in scale is taken from the science of astronomy. This example, called the dark-night-sky paradox, reveals a great deal about the scale of the universe simply by recognizing that most of the night sky appears dark and without stars.

Chapter 5 gets right to the heart of the matter of motivation. It is called "Playing for high stakes." This chapter is a smorgasbord of ideas that fit very nicely together as a feast of enthusiasms. The first part of this chapter is essentially a study of creativity using ideas gained from the young Albert Einstein, James S. Coleman, and Isaac Newton. A wide scattering of first-person anecdotes shows that children are capable of creative thinking by placing seemingly small observations into larger contexts. One of the most interesting of these stories was written by Linus Pauling, who discovered that he could reason while learning to sharpen a pencil with a knife.

The final pages are devoted to science as a career. This is intended to offer a smooth transition from the struggles of students to the equally baffling world of the creative masters.

My own dream is to introduce a new harmony to American education by helping students and teachers make their learning experiences into full-fledged scientific experiences. This is exactly what is revealed in the numerous quotations offered by the champions of science as they found themselves caught up in the great adventure. Imagine teaching any of these disciplines with the great scientists of the ages serving as your teaching assistants! They are here in this book, waiting for you.

J. W. H.

ACKNOWLEDGMENTS

A small college is sometimes looked on from the outside as analogous to a small automobile, cute but underpowered. This may be true in terms of certain types of research requiring extensive libraries and expensive laboratory equipment. However, in terms of rich conversations and interaction among students and colleagues of different specialties, the small college is a wellspring of inspiration. Creative achievement seems to be spurred by a critical mass of thinkers pushing and pulling one another along. In such an environment, individual departments are small and the critical mass comes from a coalition of disciplines. The faculty cannot share technical minutiae, so they search for common grounds on which they can all stand. Discussions have a cohesive quality that makes the writing of a cross-discipline book like this one little more than an extension of coffee time. It also makes the task of giving credit for particular contributions a very difficult one. How can I remember which teacher or student first expressed a viewpoint that, with a little twist, has now become a part of me? I can only thank my colleagues en masse for being what they are and add the hope that they savor their ideas as much in print as they did in proclamation.

Specific credit must go to my wife, Emma, for suffering patiently through so many long evenings of preoccupation. Martha Wharton, Lady d'Artagnan of all reference librarians, never failed to find the facts. Many professors offered their services. This list is by no means complete, but it does indicate the range of help that was given so freely: Dan W. Olds, physics; Dan W. Welch, physics and astronomy; B. G. Stephens and W. Scott Morrow, chemistry; E. Gibbes Patton, biology; John Pilley, James E. Seegars, and Don Scott, psychology; Dan Maultsby, sociology; Matthew Stevenson, economics; Ta Tseng Ling, government; Thomas V. Thoroughman, Joseph H. Killian, and Ross H. Bayard, history; John M. Bullard, religion; James A. Keller, philosophy; and John L. Salmon and George Adams, languages. Technical review was done with the cooperation of Raymond L. Joesten of the University of Connecticut and Mark S. Wrighton of M.I.T. Critical help in typing, proofing, and defending the English language was done by a beautiful quartet, Jody Davis, Nancy Blaisdell, Fran Scott, and Susan Dodge.

DISCOVERING SCIENCE

ONE

SCIENCE: WHAT IS THE GAME ALL ABOUT?

Scientists — that is, creative scientists — spend their lives trying to guess right.

Michael Polanyi[1]

1-1 A Few Definitions Before We Start

Before we begin, we need a bold sketch of science showing what the game is all about. The basic principles are actually quite simple. So it is appropriate to introduce them with the same sort of simplicity a child might show when using one finger to pick out a two-bar melody on the piano. Complications and refinements are added later in each of the succeeding chapters as special themes to enrich our discussion, just as counterpoints and full orchestration might be added to enhance the child's melody. A few definitions are necessary to serve as a base on which to build.

Understanding

The key idea in this book is *understanding.* A dictionary definition is of little help, for we find that understanding means comprehension and this, in turn, is defined as understanding. A better definition for our purposes emphasizes that understanding is a participation sport and therefore something of a game. *Understanding is knowing enough about something so that each part is seen in proper relationship to every other part and to the whole.* Now we know that we must be able to assemble a group of parts (that is, particular ideas) into a whole called science. This is the way that science is discovered!

Illustrations for this book are taken from many different kinds of science; all of them share this common view of understanding. The idea of *understanding* is at the very core of all scientific knowledge. That one thought is implicit in the Greek and Latin root words for sciences like geology, biology, psychology, and sociology. *Geo* (earth), *bio* (life), *psycho* (life-soul and mental processes), and *socio* (companion) define fields of *logy* (thought and discourse). Thoughtful discourse about any science requires us to discover the relationships between the whole and its parts. This concept of understanding is ancient. It is tied up in our language and yet lies all but forgotten in the press of modern times. Only better teaching and better learning will revive the idea that scientific knowledge is based on understanding.

Two Kinds of Knowledge: Rote and Cognitive

If understanding requires grasping relationships, what is *knowledge*? Dictionary definitions are quite satisfactory in this case. They indicate, among other things, that *knowledge* is acquaintance with fact. Far too many college students have the impression that rote knowledge, or memorization, gained with little or no attention to meaning, is an adequate, even praiseworthy, form of education. We will soon discover the limitations of rote learning.

Even so, there are two important, pragmatic reasons for having a substantial acquaintance with fact. The first is obvious. You cannot even begin to discuss a subject area without some sort of factual framework. In other words, you cannot begin to think without some basis for thinking. Knowledge serves as a basis for greater understanding, just as the egg precedes the omelet. The second reason for substantial acquaintance with fact is related to the physiology of the brain. Apparently, this enchanted loom maintains a series of biochemical connections among nerve cells through specialized junctions known as synapses. Memories put into the brain help to establish synaptic connections. The brain functions more efficiently if synaptic connections are used often enough to prevent the form of disruption we call forgetting. This aspect of the brain's mechanical response to rote learning accounts for the success of the nineteenth-century schoolmasters and their claims that learning disciplines the mind. We really *do* understand things better after memory patterns are established.

Knowledge begins with an acquaintance with fact, and as it becomes perfected in the brain, cognitive knowledge (or understanding) becomes possible. P. W. Bridgman, who won the Nobel Prize in 1946 for his work in the physics of high pressure, expresses the idea this way: "Intelligent teaching and learning include more than lecturing or listening or reading and remembering. The pupil must 'see' the point before he can learn, and seeing the point is something which no one can do for him."[2] Understanding or, as Bridgman puts it, "seeing the point," represents the patterns of cognitive knowledge interwoven on the brain. This is so different from the monkey-see, monkey-do character of rote learning.

The Game of Science

What, then, is the great game of *science* that we want to learn to understand? We already have most of the pieces at hand, ready to be assembled into this working definition: *science is the progressive discovery of the nature of nature.* The two most important words are *progressive* and *discovery. Progressive* indicates a continual state of improvement. The semantics of discovery are even more exciting. The root word is late Latin *discooperire* — to uncover or disclose anything previously hidden or unseen. *Discovery* is an appropriate word for the results of scientific research. Scientists do not make nature. It has been with us since the time of creation. All they can do is disclose its previously hidden secrets. A Nobel Prize is not given for following the beaten path. Therefore, when Bridgman wrote about discovery, his thoughts, like those of a combat veteran, reflected the agony of battles already fought and won. "The process of discovery cannot be put into words, for when it is put into words the discovery is over and we are dealing with history."[2] *Science,* then, is a progressive reaching out into the unknown that involves uncovering the remarkable state of nature, adding new understanding to our storehouse of knowledge, and then moving on! Science is a dynamic game and scientists get very excited about it.

Science Compared with Other Games

Familiar games like chess, bridge, tennis, football, and golf, are governed by fixed sets of rules that control all players. Winners triumph because they exhibit greater skill at the right times than the losers. Technique plays an important part in most games. Golf is a good example. The ball must be struck in such a way that it will follow a preselected path toward the hole. Until the moment of impact, the golfer must be concentrating completely on one thing: hitting the ball at exactly the right spot. If the golfer should look up in midswing, the movement changes the entire geometry of the body parts and interrelated body motions, resulting in a bad shot. Nothing in the rules of the game requires golfers to swing a certain way. The conformity and swing styles of the experts have much in common because all players must acknowledge the same physical realities. The tremendous clubhead velocity necessary to drive a ball 250 yards straight down the fairway is obtained by adding the effects of three separate pivoting units: body, arms, and wrists. Figure 1-1 illustrates the way these pivoting actions work together, adding one velocity to another.

The point of this example is to show that golf is played the same way by all participants because each of them is controlled by the same physical limitations. The rules of the game of science function in exactly the same way. They do not come from immortal decree. They exist because they seem to work better than any other options that have been tried.

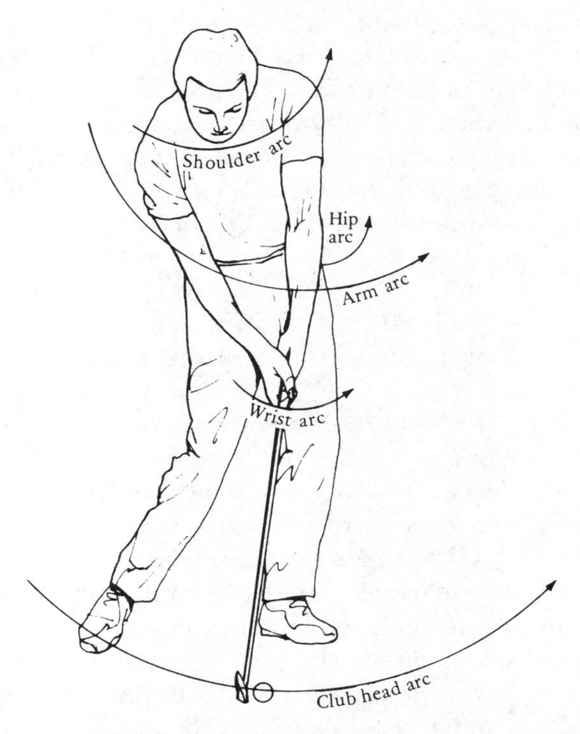

Figure 1-1 *The rules of the game of science function in the same way that physical factors control the character of a good golf swing. In both cases, the rules and methods exist because they work.*

Before we spell out the specific rules of science, we should emphasize one significant difference between this game and most others. Scientists do not play against one another except by striving for the recognition given as a reward for outstanding performances. The true competitive element in science produces a team venture directed against ignorance and lack of understanding on the part of the entire scientific community. This is easily demonstrated in the way the pronoun *we* is used in such statements as "we do not know yet. . . ." Science is a game with all of the players on one side. Free flow of information is the lifeblood of the game. Hundreds of scientific journals throughout the world provide a forum for the exchange of information. Censorship of any kind impairs scientific progress.

Modern science began to take form at the turn of the seventeenth century when Galileo Galilei (1564–1642) introduced experimentation as a means of discovering the nature of nature. We will learn more of this later in this chapter. The politically motivated censorship of Galileo during the Inquisition is an extreme example of *cultural bias.* This topic is explored in more detail later. The humanists among us may find the whole idea fascinating for it displays human emotion in conflict with scientific facts.

1-2 The First Principle of Science and Associated Assumptions

The first principle of science is a presupposition that nature can be understood. That one idea is the intellectual foundation on which all sciences rest. The key word is *can.* It implies derring-do — that is, a willingness to risk failure in the pursuit of achievement. The phrase *can be understood* is the scientist's license to practice this trade. They have built telescopes, gone to the poles of the earth, and rocketed to the moon, all in the anticipation of success. Scientists have taken on the task of searching out the state of nature because they have been heartened by the hope that it is actually possible to understand all parts of the universe. Research is the activity of hunting for hidden realities. Progress is marked by the addition of verifiable information. Every new aspect of nature that is discovered brings us closer to the goal of understanding all reality.

Assumption of Reality

Reality is our name for the state of nature. By reality we mean *that which is.* Philosophers have discussed the character of reality for several thousand years and have gained many insights without convincing each other that they know exactly what they mean when they say, *"It is."* In order to avoid unnecessary confusion and philosophical conflict, we will restrict our discussions to concrete situations where the idea of reality is easily accepted. Our task will be that of discovering nature rather than disputing its existence.

Assumption That Nature Can Be Perceived in the Same Way by Any Number of Observers

One of the other presuppositions entangled in the first principle of science is that, given the proper set of circumstances, any number of observers can see nature repeatedly in the same way. Science does not accept the view that some people have magical powers denied to others. The key ideas in this assumption are *perception* and *verification*. We have already alluded to the idea of verification by indicating that the same observations can be made over and over again by different people. Verification also involves perception and this adds a unique kind of complication. We tend to see things that our culture has taught us to expect to see. It is often difficult to break cultural traditions and see things as they really are. The concepts of verification and perception are treated in more detail further along in this chapter. The significant point at the moment is that scientists work under the assumption that anything one person can perceive, any other person can also perceive without cultural restraint.

Assumption of the Reliability of Human Reason

When we claim that nature can be understood, we imply another assumption about the *reliability* (that is, dependable accuracy) of human reason. The range of human intelligence is so great that it would be foolish for scientists to give equal weight to the reasoned conclusions of every individual. In fact, we must be honest and admit that the game of science is not democratic. We do not vote on what is right or wrong. We simply assume that our highest composite intelligence is adequate to understand nature. The assumption may or may not prove valid in the long run. We are still in the golden age of science, when reasonably intelligent individuals can still make important contributions to knowledge. Even so, we must acknowledge that the greatest breakthroughs were products of minds like those of Newton and Einstein, not just ordinary, well-trained Ph.D.'s. Perhaps a time will come when the problems to be solved will be too difficult for even the most able.

The quotation from Michael Polanyi that introduces this chapter is taken out of context, but it shows, in a rather sardonic way, how human reason overcomes its limitations. "Scientists — that is, creative scientists — spend their lives trying to guess right."[1] Guessing is an uncertain business. It is a valid part of the game if nothing better is available. Theoretical physics provides a good example. The theorist makes an estimate of what the state of nature should be. Experimentalists then test to see how accurate the estimate is. Successful guessing consists of asking the right questions and recognizing the right answers.

Assumption of Causality

The final assumption that helps to support the first principle of science is that of *causality.* This means that scientists work under the assumption that every effect has one or more causes. Understanding an action requires an equally complete understanding of its cause. In some cases this is simple; consider, for example, the recognition that rivers flow to the sea because enough rain and snow have fallen within their drainage basins. The causes of many other actions — for instance, those produced by gravity — are unknown and lie well beyond the frontiers of modern research.

The name *gravity* has been given to a force that attracts any two different masses toward one another even though they may be separated by unlimited distances. Scientists have unshakable faith that an underlying cause is responsible for such commonplace things as the way raindrops fall to the earth. Humanity has known this for more than 100,000 years. Ancient projectile points and other artifacts suggest that Neanderthals were forced to discover the behavior of falling bodies before long-range weapons became practical. A Neanderthal training session might not have been too different from passing practice on a modern football field. Some of the best minds in physics have attempted to discover the nature of gravity. The complicated history of their work is the subject of this poem.

HARD SCIENCE DOGGEREL I: THE GRAVITY SITUATION AFTER 200,000 YEARS OF RESEARCH

Neanderthals with rocks and stones,
Hunted game and broke their bones.
Testing weights and parabolics,
Are today considered Physics.

Aristotle knew that rocks and leaves,
Fall at different velocities.
And so proclaimed the different speeds,
Proportionate to their mass's needs.

Then Galileo with penduli
Discovered a hidden fallacy.
All bodies fall at equal rates,
Despite their masses and their weights.

Newton took it up from there,
And proved the Law of Distance Square.
He saw the Moon instead of tangent,
Falling downward toward our Planet.

Cavendish used a balance beam,
Balls of lead and thoughtful dream.
With these he brightly measured "G,"
Newton's constant of proportionality.

Stokes dropped spheres in a viscous fluid,
And timed the speeds for passage through it.

His Law tied back to Aristotle,
By showing why leaves slowly settle.

Then Einstein, with style and grace,
Bent the frame of Time and Space.
Gravity acts through another Law,
'Twas the path of longest time, he saw.

The web they spun, Which Science Is,
Begins to seem non-pragmatist.
Now we wait, both great and small,
For Seer to appear, Who knows it all.
And can explain this Science biz,
And truly tell us What Gravity Is.

Professors Olds and Grope

We know a great deal more about how gravity works than about why it works. But scientists do not feel that this lack of knowledge destroys the beauties of their logical system. They ascribe their ignorance to the incompleteness of their research and express absolute faith in the soundness of their exploratory methods. Scientists see the game of studying nature as having just begun and their results as incomplete.

It is time to go back and run through the melody we have been picking out with one finger. The whole game of science rests on these four presuppositions:

1. Nature, whose reality we can substantiate, can be understood.
2. Given the proper set of circumstances, nature can be perceived in the same way by any number of observers.
3. Human reason is reliable and adequate to understand nature.
4. Every effect has a cause.

The argument that supports the use of the standard golf swing also supports the scientist's right to use these four assumptions. They work! Playing the game in this manner has been successful. We have developed a well-verified body of knowledge about nature that we can offer as proof.

First Principle of Science Contrasted with the First Principles of Other Disciplines

There is no denying that fundamental ambiguities lie hidden in the roots of supposedly exact sciences. We also find ambiguities if we examine the presuppositions on which other intellectual disciplines are based.

The first principle of education seems to be the presupposition that learning is possible. Note how we have shifted the responsibility for learning from the faculty to the student. We have not said that teaching is possible but that learning is possible. P. W. Bridgman goes even further by explaining the mechanism of learning in his remark, "The pupil must 'see' the point before he can learn."[2]

The first principle of law appears to be the presupposition that society operates best under a fixed set of rules. These rules come from four sources: the authority of custom, prior judicial decision, legislative statute, and decrees of government agencies. The first two are conservative in nature and tend to hold society on the course of the past. The last two may be more innovative. At the international level, there is nothing in the first principle of law that dictates the kinds of rules that a society may choose for itself. Under these conditions, there should be little wonder that different societies find themselves in conflict with each other across international boundaries even though their systems of law are based on the same first principle.

The first principle of democracy is apparently the presupposition that the most desirable form of government is the one that produces the most good for all the people. The key words *one, produces, most good,* and *all* sound substantial, but they are subject to various interpretations. Governments have collapsed as a result of conflicting views of what these words should mean. Conflicts are almost inevitable if the presupposition is twisted to mean "the most good for the most people." Minorities may reject the idea with revolutionary results, as in 1776, for example.

We may even examine the first principles on which the two great types of religions are based. Theistic religions, such as Christianity, acknowledge a god. They seem to be based on the presupposition that there is a mutual concern between their gods and mankind. Atheistic religions acknowledge no god and are based on the presupposition that man has an ultimate concern. Paul Tillich considered *ultimate concern* to mean devoted, unconditional seriousness on the part of every individual involved.[3] Where is faith? It is unjust to restrict the concept to a line, but if forced to do so, we might imagine that faith begins as a trust in the validity of the presuppositions.

Faith of this sort is shared by thinkers in all disciplines. Scientists, teachers, lawyers, politicians, political theorists, theologians, and philosophers use faith as a license to develop their thoughts and play the game.

1-3 Some Components of the Game of Science

Perception

Perception is defined as the direct acquaintance with anything through any of the five senses: sight, hearing, touch, taste and smell. Perceptive abilities vary greatly from individual to individual over an enormous range of values. Some of us have perfect vision; others are near-sighted or far-sighted. Some of us are deaf to high frequency sounds that others can hear with ease. Some of us wince at the thought of slight pains while others with very high pain thresholds seem to be able to tolerate almost anything. Our concern is not with differences imposed by variations in the structure of the human body as much as it is with a lack of desire to perceive details.

Most of us can see leaves when we look for leaves, but not when we look for trees. We tend to see the outlines and shapes of objects rather than their component parts. Perception of details is a skill that depends in part on a physical principle known as resolving power.

Concepts of Resolving Power and Pattern Recognition The *resolving power* of an optical instrument, such as a telescope or a microscope, is the measurement of its ability to distinguish clearly between two different things. Anton van Leeuwenhoek (1632–1723) is listed in the *Biographical Dictionary of Outstanding*

Men because he was the first to arrange simple lenses as a primitive microscope and call the attention of the Royal Society of London to living forms in a drop of water. Up to that moment, his usefulness to society had been as a family man and custodian. Leeuwenhoek's invention helped the human eye to distinguish clearly among living protozoa, very small single cell animals. This was a simple beginning that led to more sophisticated generations of instruments and incredible advances in science and medicine.

By Leeuwenhoek's time, Galileo had already developed a 30-power telescope, discovered the moons of Jupiter and forty

Figure 1-2 *Galileo (1564–1642) was an experimental physicist. He worked with pendulums having different masses and proved that Aristotle was wrong in believing that objects fall to the earth at velocities proportional to their masses. (Courtesy, American Museum of Natural History)*

stars in the Pleiades, and found that the Milky Way is made of a mass of individual stars. All of these discoveries were made possible by the resolving power of the new optical technology. The prescientific society of that time was ill-prepared for the enormous consequences of these facts. Nicolaus Copernicus (1473–1543) had already destroyed the dogmatic viewpoint that the earth was in the center of the universe. He demonstrated that our world is simply a satellite revolving about the sun. Three and a half centuries later we are still learning the full meaning of this lesson. Nature is not ours to dominate and exploit. We must struggle continually to maintain the right to ensure a livable environment for ourselves and our descendants on this conspicuously small, spinning planet. Ideas have resolving power, too!

Despite advances in instrumentation, modern science ultimately remains dependent on the perception and resolving power of the five senses. There are no other known avenues for passing information to the brain.

The resolving power of the brain functions in part through a phenomenon known as *pattern recognition.* Some kinds of pattern recognition are so automatic that we hardly notice what we are doing. Reading is a good example. The letters on this page are nothing more than patterns in black and white combined in a standardized way to represent the sounds of spoken language. The idea is a basiç part of Western civilization. The same symbols transfer easily from English, au Français, al Español, ins Deutsche, aut in Latinum. Civilization must be reborn with each new generation as children spend the first few years of their formal education learning the technique of this kind of pattern recognition. We read in a manner similar to the way we drive our automobiles, by granting significance to the positions and patterns of things in space.

Pattern recognition is also a crucial part of all sciences. The point is well illustrated in Chapter 2 by the pioneering work of Leonardo da Vinci, Charles Darwin, and Alfred Wegener. Each of these men saw a critical set of patterns in the array of scientific facts known to them.

Most people are not capable of producing major triumphs on the scale of da Vinci and Darwin, but we all have some ability to search for and recognize visual patterns. The truth of this is revealed in a common folk saying, "I can't remember names, but I never forget a face."

The human brain seems to be much more successful at recognizing visual patterns than any of the present generation of electronic computers. Our abilities to read distorted handwriting illustrate the point. Modern computers are unable to read the wide range of handwriting styles that schoolteachers deal with routinely when grading papers. Apparently the brain is able to construct basic *models* to serve as common denominators against which the eye measures each writing style. We may miss a word or two but we can generally make out the ideas by inserting words of our own to flesh out the blank spaces. Computers are still too narrow-minded for this tactic.

Concept of a Scientifically Verified Fact Facts are the products of perception and are therefore as complicated as perception itself. Complicated or not, we need to understand the nature of facts because they are the building blocks of science. One useful definition is given in the *Encyclopedia of Philosophy and Psychology:* "A fact is an objective datum (reference point) of experience." An objective experience is one that contains no trace of personal feeling. It cannot be influenced by an individual's emotions or assumptions. Snatches of two actual conversations with an art gallery guard and a mathematics professor put these ideas in much simpler words.

The guard was asked to define a fact. His answer was emphatic. "A fact is a sure thing!" The mathematician was then asked to define a sure thing. "A sure thing is an event that will always occur. It has a statistical probability of occurrence of one. This means the same thing happens every time." The art gallery guard nodded his head in agreement and said, "You can bet on it!"

Put all of this together and we can agree that *a fact is a sure thing that we can bet on because we have had an objective experience with it.* The key words are: *fact, sure* and *objective experience.*

Much of the cohesiveness of scientific thinking is due to a general agreement about the nature of the facts that scientists know and work with. This cohesiveness is almost inevitable. Most scientists have shared the same kinds of experiences. They have taken the

Figure 1-3 *Cultural bias may prevent you from seeing that the lump of material floating in this glass of water is a rock. It is a piece of pumice, a volcanic rock containing many holes produced by steam escaping during its formation.*

same courses in college, listened to the same sorts of lectures, and performed similar laboratory experiments. Scientists compose a social subculture distinguished by a common set of values, attitudes, and operating procedures.

No single investigator has the time to duplicate every experience and verify every fact recorded in the scientific literature. Fortunately, this is not necessary because the rules of the game of science require that facts *must be verified* before they are accepted for publication. Researchers know that their scientific reputations depend on the accuracy of their reports. Errors will be discovered and exposed by skeptical colleagues if they are unable to repeat the experiments or verify the observations. Once the facts are known to be accurate, they become a part of the accepted view of nature and are taught to succeeding generations of students.

Cultural Biases, Our Learned and Shared Prejudices *Cultural biases* are limitations imposed on the creative aspects of science, for they prevent us from grasping the meanings of our data. It is easy to show how cultural bias works. Figure 1-3 is a picture of a lump of something floating in water. It is a rock! Surprised? We expect rocks to sink in water, not float. Most rocks do sink, but this is a rare type called pumice. This light-colored glassy lava is composed chiefly of five chemical elements: potassium, sodium, aluminum, silicon, and oxygen. The most outstanding characteristic of this strange rock is its cellular structure, which resembles the texture of bread. The holes were made by expanding steam escaping from the once liquid lava. As a result, pumice is less dense than water and will float like a cork.

Cultural bias creates a way of looking at things in terms of what we already believe to be true. Effective learning is frequently accomplished by setting biases aside, looking at information in a new way, and making a personal discovery of what really is true.

Intellectual history, particularly the history of science, is a study of the ways in which cultural biases have operated through time. We will see several examples in Chapters 2 and 3. Individuals who make scientific discoveries and recognize their meanings leap over the cultural biases of their own times and literally become part of the future. To do this they must ignore cultural biases that hold back their contemporaries. Old ideas are declared invalid because the discoverers have learned to see nature in new ways. After each major breakthrough there is usually some technological advance that lends support to the new idea and improves its chances for general acceptance.

Charles Darwin's (1809–1882) revolutionary book *On the Origin of Species,* published in 1859, was more fully appreciated by geologists and biologists than by the general public. These scientists understood that the progressive change of living things through time by a process Darwin called *natural selection* was virtually inevitable on an earth of great antiquity. They also knew that the earth was very old; that fact had been known for seventy years. Dr.

Figure 1-4 *Charles Darwin (1809–1882) published his revolutionary book* On the Origin of Species *in 1859. In it he set out a theory of organic evolution through the processes of natural selection that occur over vast spans of geological time. (Courtesy of the Library of Congress)*

James Hutton of Edinburgh had published a book, *Theory of the Earth,* in 1788 that presented compelling evidence of the great length of geologic time. In 1831, Sir Charles Lyell (pictured in the next chapter, Figure 2-3) improved on Hutton's work and offered even more compelling arguments in a three-volume treatise

Figure 1-5 *James Hutton (1726–1797) published* Theory of the Earth *in 1788. The last line quoted below set the stage for Darwin, for it suggested the vastness of geologic time. "The result, therefore, of our present enquiry is, that we find no vestige of a beginning, — no prospect of an end."*[4] *(Courtesy, American Museum of Natural History)*

entitled *Principles of Geology*.. Darwin's theory was consistent with the knowledge that there was ample time for evolutionary changes to take place. Much of the general public felt quite differently.

For several centuries the age of the earth had been fixed by a popular belief that the creation as described in Genesis had occurred in 4004 B.C. Biblical scholars had established this figure by interpreting the "begats." Lingering remnants of this polarization between Biblical creationists and the scientific community still exist today. All of us function within constraints imposed by our cultural biases. Unfortunately, cultural bias seems to be limited by, and no broader than, our experience or lack of it.

Even James Hutton did not break out of the cultural biases of his time without some help from earlier thinkers. A selection from Samuel Pepys's diary for May 23, 1661, gives us a glimpse of the kinds of conversations that scientific thinkers were having at that time. Pepys served as president of the Royal Society when Isaac Newton published the *Principia.* Jonas Moore, Ordnance General under Charles the Second, was responsible for engineering work in East Anglia, where he used Dutch engineers in a project to drain the Fens. The Dutch engineers taught Moore, he taught Pepys, and the bias began to collapse.

> . . . To the Rhenish wine house, and there came Jonas Moore, the mathematician to us; and there he did by discourse make us fully believe that England and France were once the same continent, by very good arguments, and spoke of many things, not so much to prove the Scripture false as that the time therein is not well understood.[5]

Models

Most of us have learned to think in terms of models. We learn to find our way through strange cities and across the country by using maps. Maps are just models of the real countryside, drawn to a small scale for handy reference. The word *model* is derived from a Latin word, *modulus,* to mean a small measure. For scientific purposes we may define *model* to mean a simplified representation of a real object, system, or operation. Figure 1-1 is a model of a golf swing, not a real three-dimensional swing having such properties as momentum. Models are useful because they permit us to analyze components of a larger system. This entire chapter is a model depicting the game of science. The book contains many additional models, each used as parts of larger ones. Together, they furnish us with an approach to understanding science.

Logical Systems as Models The two broad forms of logical thought, *inductive and deductive reasoning,* are used constantly in everyday life. Even though these two forms of thought are practiced by all of us regularly, many of us would probably have some difficulty explaining the fine points of each type of reasoning and making appropriate distinctions between them. The situation is something like that of walking. We are better at doing it than explaining exactly how it is accomplished.

Deduction We will consider *deductive reasoning* first because it is the type of reasoning we use most often. We reason deductively when we infer a conclusion about the meaning of some set of information by using what we believe to be the principles that govern such things. The logical thinking demanded by deduction is an act of classifying data to determine which principles apply in a specific case. The entire body of standard legal thinking is deductive in nature. Legal arguments by attorneys are simply presentations of the information and opinions for each side. It is the judge who must decide which statutes and prior adjudications cover the case.

A more personal example of deductive thinking is provided by the decision to go on a diet to lose weight at a given rate. Authorities indicate that each extra pound of human fat represents the consumption of about 3,500 excess calories. Therefore, if we decide to lose a pound a week it is necessary to restrict our normal diet by 500 calories a day. In deciding which foods to eliminate from our diet and which foods to keep, we must make further deductions. The first step is to translate food types and quantities into calories. For example, a cup of peanuts contains 840 calories. Desserts average from 300 to 550 calories. One fancy cookie may contain as many as 50 calories. The second step is to find out what kinds of amounts of the different foods constitute a balanced nutritious diet. The final step is the decisive one. Our choices must be made from a range of options. The best choices represent deductive thinking that recognizes a responsibility to maintain our own health.

Interestingly enough this idea of responsibility is rooted in our language, much of which has evolved through the same deductive processes. The modern English word *diet* may be traced back through Middle English and Old French to the Latin *diaeta* and the Greek *diaitia,* both of which mean diet and mode of life. *Ai* is an even older Indo-European root of these words. It means to give or to allot. The same root is found in the Greek *aitia,* indicating cause or responsibility. Thus, *diaitia* includes the ideas of cause and responsibility, two ideas familiar to anyone who has tried to diet.

The very act of dieting requires that will power overcome "gut power." Will power stems from a certainty that the brain has made the properly reasoned (in this case deductively reasoned) decision on the basis of all the important facts. No wonder the brain is the most important of all our organs. Our health depends on the brain's ability to stay in control of the body.

Induction and the Principle of Least Astonishment Inductive reasoning has taken place when a collection of facts is judged to imply that some new conclusion is probably correct. The inductive process admits the chance of error. There is a great gulf between the deductive and inductive processes. If facts are classified correctly, deductive conclusions drawn from them are as certain as the facts themselves. Inductive reasoning breaks new ground and contains an element of uncertainty. *Inductive reasoning* is the logical thinking process that takes a set of facts and draws from them some new model. Logicians may call the new model a hypothesis, a theory, or just a generalization. It is a conclusion, perhaps just a guess, about the state of nature indicated by the supporting facts.

We must find out how inductive inferences are drawn. The basic expression is simple enough:

$$\text{Fact } 1 + \text{fact } 2 + \text{fact } 3 + \ldots + \text{fact } n \xrightarrow{\text{Implies by least astonishment}} \text{Conclusion(s)}$$

A working definition of the principle of *least astonishment* may be stated this way. Least astonishment is a mechanism for selecting the most probable conclusion from among a number of competing options on the basis of minimum surprise. All the others are eliminated because they seem too astounding to be believable. The conclusion chosen through the principle of least astonishment is accepted as the best available in light of all that is known.

An excerpt from *The Study In Scarlet,* the first of the Sherlock Holmes stories by A. Conan Doyle, illustrates the principle of least astonishment and how it operates.[6] Dr. Watson, who has just met Holmes, is a little doubtful about the reasoning abilities of this strange man. The scene opens at 221-B Baker Street, London. Holmes and Watson are looking out of the window of the apartment they have just begun to share. Watson remarks:

> "I wonder what that fellow is looking for?" I asked, pointing to a stalwart, plainly dressed individual who was walking slowly down the other side of the street, looking anxiously at the numbers. He had a large blue envelope in his hand, and was evidently the bearer of a message.
>
> "You mean the retired sergeant of marines," said Sherlock Holmes.
>
> "Brag and Bounce!" thought I to myself. "He knows that I cannot verify his guess."
>
> The thought had hardly passed through my mind when the man whom we were watching caught sight of the number on our door and ran rapidly across the roadway. We heard a loud knock, a deep voice below, and heavy steps ascending the stair.
>
> "For Mr. Sherlock Holmes," he said, stepping into the room and handing my friend the letter.
>
> Here was an opportunity of taking the conceit out of him. He little thought of this when he made that random shot. "May I ask, my lad," I said, blandly, "what your trade may be?"
>
> "Commissionnaire, sir," he said, gruffly. "Uniform away for repairs."
>
> "And you were?" I asked, with a slightly malicious glance at my companion.
>
> "A sergeant, sir, Royal Marine Light Infantry, sir. No answer? Right, sir."
>
> He clicked his heels together, raised his hand in salute, and was gone. . . .
>
> "How in the world did you deduce that?" I asked. [Conan Doyle made an error there. Sherlock Holmes reasoned inductively, not deductively.]
>
> "It was easier to know it than to explain why I know it. If you were asked to prove that two and two made four, you might find some difficulty, and yet you are quite sure of the fact. Even across the street I could see a great, blue anchor tattooed on the back of the fellow's hand.

That smacked of the sea. He had a military carriage, however, and the regulation side-whiskers. There we have the marine. He was a man with some amount of self-importance and a certain air of command. You must have observed the way in which he held his head and swung his cane. A steady, respectable, middle-aged man, too, on the face of him — all facts which led me to believe he had been a sergeant."

According to the presupposition of causality, every result has a cause. The principle of least astonishment offers a rational way of deciding how to pick the best explanation from a set of possible ones. Selection is based on, and then justified by, what seems to be least astonishing in light of what is already known about such things. At best, this principle is a way to make an educated guess, as Michael Polanyi recognized in his statement about the way "Scientists . . . spend their lives in trying to guess right." Philosophers have other names for the principle of least astonishment — among them, *logical simplicity, parsimony,* and *Occam's razor.* Least astonishment is just a facet of the logician's concept of *probabilism,* the doctrine that certainty is impossible but that probability suffices to govern belief and action.

How Modern Science Can Stand Successfully on the Principle of Least Astonishment Modern science is a successful venture. It may seem odd at first that the structure could be held together by nothing more rigorous than a series of educated guesses. The answer lies in the way these guesses are used. Scientists consider each new conclusion about the probable state of nature to be nothing more than a prediction. Verification tests are necessary to determine how true the prediction is. The theoretical and experimental branches of science share this role of thinking up predictions and testing them. A new conclusion becomes part of the larger fabric of knowledge *only after two crucial criteria are satisfied.* The first criterion is that nothing more probable is discovered to replace the conclusion. The second is that no contradictions are found that require the conclusion to be rejected. In short, a new conclusion is accepted after the scientific community has seen that it works. The system of expanding scientific knowledge is not unlike that used by a sculptor chipping away at a block of marble to release the form he or she imagines is inside. Scientists presuppose that they will find reality by making guesses and then abandoning those that become untenable. The guesses that remain either must contain reality or must eventually be discarded through the crisis of contradiction. This is a major point in Chapter 2.

Hypotheses and Laws If the principle of least astonishment is used properly, well-verified conclusions and working models may be considered reasonably accurate pictures of nature. If they are defendable, conclusions pass from the *hypothesis* category to the status of laws. Familiar examples such as Newton's law of gravity, Ohm's law, Boyle's law, Charles's law, and Hooke's law are not final truths. They are *probabalistic statements,* elevated to the status of laws because no cases of contradiction are known within accepted limits. Sometimes surprising things are learned about the limits that must be imposed on the laws of science.

Consider, for example, Einstein's development of relativistic mechanics in the early 1900s. He found that Newton's ideas did not apply at velocities approaching the speed of light. Einstein's insights, supported by tests performed with instruments and resolving powers unknown in Newton's time, represented a crisis of contradiction (this idea will be discussed more fully in Chapter 2). Newton's laws of motion and Newtonian mechanics had to be redefined as a special case that described behavior at low velocities. We now have two sets of mechanics, each discovered through the principle of least astonishment. Our view of nature is more accurate because the crisis of contradiction forced us to see nature in a different way.

The Fabric of Science Models are used in science to represent a simplified version of a real object, system, or operation. Figure 1-6 is a model of the way all the facts and conclusions of science are woven together to form a single cohesive fabric.

The bottom line consists of a series of facts representing the lowest order of complexity. The web of lines just above these basic facts represents the logical ties of least astonishment, leading to a set of conclusions. These in turn become facts supporting a more complex web of logical ties and higher levels of conclusions. Scien-

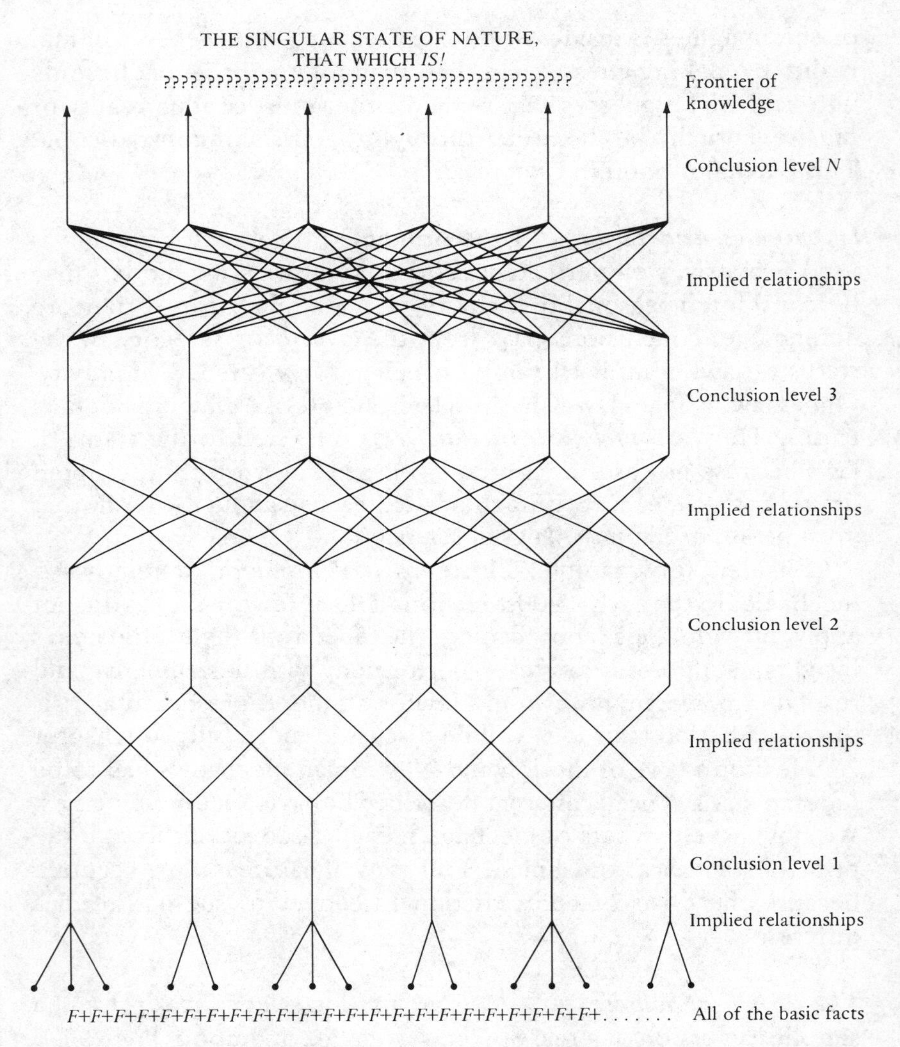

Figure 1-6 THE FABRIC OF SCIENCE *This device illustrates the logical pattern of the fabric, woven out of the implied relationships between facts and conclusions. The ultimate level of achievement is the discovery of that which "is." (From John W. Harrington,* To See a World, *St. Louis: The C.V. Mosby Company, 1973)*

tists are weaving the fabric toward the final goal of scientific research, the discovery of the complete state of nature.

Scientists play the game of extending and improving this fabric.

Figure 1-7 *The first few bars of "Morning Mood," the first movement of Edvard Grieg's* Peer Gynt *suite no. 1, are a good illustration of an imitative abstraction. Science is like this, a model of reality with a life of its own.*

In this first chapter, with a simple one-finger melody, we have laid out all the rules. But we have not yet answered the most important questions: How do you win the game? How are discoveries made? These are exciting questions. In answering them we will try to develop an appreciation for hard science and for the people who have created its fabric. We will discover that science is an abstraction, a model of reality with a life of its own. Knowledge grows as an imitative counterpart of some bit of nature. We must learn to appreciate the imitations in order to appreciate the larger fabric.

"Morning Mood," the first movement of Edvard Grieg's *Peer Gynt* suite no. 1, is a good illustration of things to come. Grieg has modeled place, action, time, and emotion in just a few bars of this beautiful abstraction. Our task is to discover the realities of nature revealed by the singing flutes.

Annotated References

1. Michael Polanyi, *Personal Knowledge* (New York: Harper and Row, Harper Torchbooks, The Academy Library, 1958), p. 143.
2. P. W. Bridgman, *The Way Things Are* (Cambridge, Mass.: Harvard University Press, 1959), pp. 77–78.
3. Paul Tillich, "First Dialogue," in *Ultimate Concern: Tillich in Dialogue,* ed. D. Mackenzie Brown (New York: Harper and Row, 1965), pp. 7–16.
4. James Hutton, "Theory of the Earth," in *Contributions to the History of Geology,* ed. George W. White (Darien, Conn.: Hafner Publishing Co., 1970), V, 304.
5. *Everybody's Pepys: The Diary of Samuel Pepys 1660–1669,* ed. O. F. Morshead (New York: Harcourt Brace and Company, 1926), pp. 93–94.

6. Arthur Conan Doyle, "The Study in Scarlet," in *The Boy's Sherlock Holmes,* ed. Howard Haycraft (New York: Harper and Brothers, 1892), pp. 20–21.
   Doyle frequently uses this method of identifying the ex-soldier by bearing and suntan. "The Case of the Greek Interpreter" is an especially notable Holmes story, for both Sherlock and his even more masterful brother, Mycroft, get into the act.

TWO

WINNING THE GAME OF SCIENCE: FROM THE CRISIS OF CONTRADICTION TO DISCOVERY

To know ourselves diseased
is half our cure.

Alexander Pope

2-1 Some Characteristics of Creative Behavior

Creativity, the Joke Analogy, Passive and Active Behavior

Intelligence may be defined as the ability to perceive relationships. *Scientific creative intelligence* is a special aspect of general intelligence, demonstrated by the ability to see a set of data and draw new scientific conclusions from it. We all have this ability to some degree, although the amount varies from individual to individual. The capacity for creative intelligence may be demonstrated by a person's ability to grasp a joke.

A good raconteur presents a story that defines a situation and the way we are to look at it. These are the straight lines. Then, immediately after a short pause signals that we must get ready to use our creative intelligence, the punch line is sprung. At this instant we must shift from the role of passive listeners to that of active participants. The punch line shows us how to see the situation in an entirely different way. Our joyful, triumphant feeling of success is marked by a chuckling laugh as creativity occurs. The joke falls flat if we do not make the proper creative leap. This kind of creativity began with a shift from a passive to an active role and ended with an acknowledging chuckle.

Here is an example to demonstrate the point. Note how easily the creative moment is recognized.

> Two strangers, a man and a middle-aged woman, were sitting beside one another on an airplane. The woman had a magnificent diamond ring that flashed brilliantly every time she moved her hand. The man was fascinated. Finally, unable to contain his curiosity any longer, he said, "That's the most magnificent diamond I've ever seen. It must be very famous. Is there a story attached to it?"
>
> "Oh yes, this is the Klopman diamond. It's very famous, but there's a curse attached to it."
>
> "A curse, that's dreadful. What is it?"
>
> "Mister Klopman."

Cognitive Learning and the Active Role

We have already developed the concept of understanding as an important aspect of learning. Understanding enables us to see the parts in proper relationship to each other and to the whole. Cognitive learning is a creative act performed by identifying the parts and drawing our own conclusions from them. This type of learning begins by assuming an active role and forcing ourselves to see the collected facts in new ways. It is easy to demonstrate how this is done.

Imagine that we are standing on the bank of a river watching it flow past us to the sea. Someone asks, "Why does the river flow this way?" The obvious answer is that the riverbed slopes to the sea and gravity pulls the water down that way. Oh! but gravity is a force directed straight down toward the center of the earth, not toward the ocean 200 kilometers away. That is true. Perhaps we should break down the problem into smaller parts and see what takes place.

What would happen to a single unconfined cubic meter of water if it were placed on a sloping surface? It is difficult to imagine such a block standing for more than a moment without collapsing in all directions, but moving chiefly downhill. We may expect this to happen, for the fluid character of water is well known. In more scientific terms, the water has a low internal shearing strength. The molecular structure partially bonds many units together in strings that slip past one another as the change of shape takes place. Friction between them is called viscosity, and a cold stiff fluid like molasses in January would change shape more slowly than water because its viscosity, or internal shearing strength, is much greater.

The river that we have been watching may be represented by a long line of 200,000 water blocks set one behind the other. The one in front of us cannot collapse until the one just downstream has made room for it. As each water block moves or enters the ocean, it makes room for a proportionate movement of each block upstream. It is easy to see that a flowing river is just a long line of collapsing water blocks with movement taking place across the channel as well as downstream. Eddies and crosscurrents develop where some lines of water blocks move faster than others, due to differences in friction along the irregular bottom and banks.

If you are visualizing this model for the first time, you may feel a little burst of gleeful joy. This feeling results from assuming the active role by making a creative connection between a real river and this mechanical model. Can you make a similar model to explain why the ocean basins do not spill over from the daily addition of river water? (The answer is that water is taken from the sea by evaporation, with rain, snow, and river flow maintaining the balance.)

2-2 The Crisis of Contradiction Followed by Discovery

Normal Science

In his book *The Structure of Scientific Revolutions,* Thomas S. Kuhn has made an exciting study of the mechanics of discovery.[1] He says that most scientists are engaged in problem solving, using the techniques and concepts of their day. He calls this "normal science" and defines it as ". . . research firmly based upon one or more past scientific achievements, achievements that some particular scientific community acknowledges for a time as supplying the foundation for its further practice." Our model of the fabric of science (Figure 1-6) essentially represents Kuhn's normal science. The mechanics of problem solving are those of collecting facts and reasoning a conclusion from them through the principle of least astonishment:

$$\text{Fact 1} + \text{fact 2} + \text{fact 3} + \ldots + \text{fact } n \xrightarrow{\text{Implies by least astonishment}} \text{Conclusion(s)}$$

The limitations of normal science are obvious. Refinements and some extensions can be made in the fabric, but no startlingly new aspects of nature can be discovered in this way. The limitations on creative thought imposed by the principle of least astonishment are too prohibitive.

The Role of the Anomaly: A Crisis of Contradiction

Now suppose that routine research produces some startling new fact that completely contradicts some previously held conclusion. A fact like this is an *anomaly,* a deviation from the expected state of things. Anomalies exist only in our inaccurate models of nature. There can be no anomalies in nature. Nature cannot be surprised, but scientists and students can. A great new anomaly is always something of a shock to the established order of things. The rules of the game are such that every real anomaly must be dealt with. Intellectual honesty prevents scientists from being satisfied with pretending that the fact is simply not there. The reasoning expression now takes on this new form, indicating a crisis of contradiction:

$$\text{Fact 1} + \text{fact 2} + \text{fact 3} + \text{fact 4} + \text{fact anomalous} \xrightarrow{\text{Implies by least astonishment}} ?$$

A valid crisis of contradiction marks a healthy stage in the development of science: as Alexander Pope said, "To know ourselves diseased is half our cure." We now know that our old beliefs were somehow in error and that new ones must be discovered.

Kuhn examined a number of classical crises that have occurred in the history of science and worked out the pattern of changes that followed. We will fit a discussion together in his own words by taking a few short quotations from their proper context and sequence. Some slight changes in noun and verb forms have been made to improve the grammatical fit as these quotations are strung together. Kuhn's genius deserves to be seen in the flesh rather than through ghostly paraphrased accounts.

What is a Paradigm?

Kuhn discloses a process of uprooting the once falsely happy normal science and creating by revolution an entirely different fabric in which the new normal science may flourish. He considers the scientific paradigm to be the unit that changes. The concept of the scientific paradigm, what it is and how it functions, has caused more controversy among philosophers and historians of science than any other part of Kuhn's thesis. His briefest definition of the paradigm is: "a universally recognized scientific achievement that for a time provides model problems and solutions to a community of practitioners."[1] The concept is simple enough if we think of the paradigm as a view or model of nature held and used by some group of scientists. This view may be a grand compilation including many smaller views. James Hutton's *Theory of the Earth* (1788), Charles Lyell's *Principles of Geology* (1830), and Charles Darwin's *On the Origin of Species* (1859) are all important examples of grand paradigms that have controlled the path of scientific history.

Students are rarely on the frontiers of research, but they are constantly involved in extending the frontiers of their own learning. Paradigms show us how to extend our thoughts into new areas. We use them to think things through and to place new observations into a larger framework. Note how easy it is to follow through a complex set of facts using a single paradigm as the connective thread.

John Wesley Powell (1834–1902) first published the concept of *base levels of erosion* in 1875. Clarence E. Dutton (1841–1901) refined the idea six years later. Both men recognized that streams could not erode their valleys any deeper than the basins into which drainage occurred. Sea level is the ultimate base level of erosion. Of course, there are many temporary base levels imposed by resistant rocks that can block stream action for a time. The limestone cap that impedes erosion at Niagara Falls is a resistant rock maintaining a temporary base level.

Think of waterfalls as marking temporary base levels, restricting, for a time, deep erosion of the land lying immediately upstream. Think, too, of the theoretically inevitable day when all temporary base levels have been removed and a stream has cut down its valley nearly to sea level. The Mississippi River has achieved this success between the Gulf of Mexico and Cairo, Illinois. This part of the river gradient is a long low slope approaching sea level as a limit. Think, too, of the great plains of the world. We find them on every continent except Antarctica, which is buried in ice. To some degree, these plains owe their forms to the gradients of the rivers flowing

Figure 2-1 *John Wesley Powell (1834–1902) led the first geological surveying party down the Colorado River and through the Grand Canyon. While serving as second director of the United States Geological Survey, he brought order to the homesteading of the West by providing Congress with scientific judgment. (Courtesy, United States Geological Survey)*

across them to the sea. The Powell-Dutton paradigm helps us understand why plains are the fundamental land form, why most people in the world live on plains, and why the average elevation of continents is only about 800 meters (a half mile).

The Structure of Scientific Revolutions: Discovery

Kuhn thinks of every paradigm in science as having once been new. Here are the common qualities that make them revolutionary:

> They served for a time implicitly to define the legitimate problems and methods of a research field for succeeding generations of practitioners. They were able to do so because they shared two essential characteristics. Their achievement was sufficiently unprecedented to attract an enduring group of adherents away from competing modes of scientific activity. Simultaneously, it was sufficiently open-ended to leave all sorts of problems for the refined group of practitioners to resolve.[1]

Here is his picture of the crisis and the events that follow:

> Discovery commences with the awareness of anomaly, i.e., with the recognition that nature has somehow violated the paradigm-induced expectations that govern normal science. It then continues with a more or less extended exploration of the area of anomaly. And it closes only when the paradigm theory has been adjusted so that the anomalous has become the expected. Assimilating a new sort of fact demands a more than additive adjustment of theory, and until that adjustment is completed — until the scientist has learned to *see* nature in a different way — the new fact is not quite a scientific fact at all.[1]

Who Wins the Game?

> The new paradigm, or a sufficient hint to permit later articulation, emerges all at once, sometimes in the middle of the night, in the mind of a man deeply immersed in crisis. What the nature of the final stage is — how an individual invents (or finds he has invented) a new way of giving order to data now all assembled — must here remain inscrutable and may be permanently so. Let us here note only one thing about it. Almost always the men who achieve these fundamental inventions of a new paradigm have been either very young or very new to the field whose paradigm they change. And perhaps that point need not have been made explicit, for obviously these are the men who being little committed by prior practice to the traditional rules of normal science,

are particularly likely to see that those rules no longer define a playable game and to conceive another set that can replace them. The resulting transition to a new paradigm is scientific revolution. . . .[1]

Chronological Sequence of Events. Winning the game may be entirely represented in chronological sequence:

1. Problem solving of normal science in the period of time before the old paradigm is questioned:

$$\text{Fact 1} + \text{fact 2} + \text{fact 3} + \ldots + \text{fact } n \xrightarrow{\text{Implies by least astonishment using old paradigm}} \text{Conclusion(s)}$$

2. Normal science brought to a halt by the crisis of contradiction as the old paradigm is challenged:

$$\text{Fact 1} + \text{fact 2} + \text{fact 3} + \ldots + \text{fact } n + \text{fact anomalous} \xrightarrow{\text{Implies by least astonishment}} ?$$

3. Revolution! The anomaly disappears because under the new revolutionary paradigm it becomes a predictable fact:

$$\text{Fact 1} + \text{fact 2} + \text{fact 3} + \ldots + \text{fact } n \xrightarrow{\text{Implies by least astonishment}} \text{New paradigm}$$

4. Problem solving of normal science resumed in the period of time after the revolution:

$$\text{Fact 1} + \text{fact 2} + \text{fact 3} + \ldots + \text{fact } n \xrightarrow{\text{Implies by least astonishment using new paradigm}} \text{Conclusion(s)}$$

The fourth step, in which normal science and problem solving resume, requires general acceptance by the scientific community. This is the stage that Kuhn feels must have two essential characteristics. The new paradigm must attract an enduring group of adherents away from competing activity. It must also be sufficiently open ended to leave problems for the new practitioners to solve.

2-3 Case Histories of Scientific Revolutions via the Crisis of Contradiction

Classic illustrations of scientific revolutions exist in the work of Nicolaus Copernicus (1473–1543) and Galileo (1564–1642). Both men recognized anomalous facts and used them as bases for revolutionary new concepts. Archimedes (287–212 B.C.) is supposed to have said, "I could move the earth with a lever, if I had a place to stand." In a way, Copernicus did move the earth. He displaced it from an imaginary position at the center of the universe to that of a minor planet orbiting the sun with all the other planets. The human species has never been sure of its own place and importance in nature since Copernicus published *On the Revolution of the Celestial Orbs* (1543).

Galileo was one of the scientific practitioners who was attracted to the new Copernican paradigm. His own observations of the moons of Jupiter supported the Copernican thesis, for he had seen them revolve around their mother planet. He, too, wrote a book on the topic. However, Galileo's place in science stands on a different kind of greatness. He devised experiments to discover how things happen. His work on falling bodies was particularly important and experimental physics was founded on it.

The science that Galileo practiced is quite different from a science that studies the earth. Geology is an historical science. Geologists study occurrences that have already taken place, very few of which can be studied experimentally because they cannot be reproduced. Extinct dinosaurs cannot be brought back to life so that we may observe their evolution. Seas cannot be restored across a continent. The continents cannot be rejoined and moved back to the various parts of the earth they occupied over the last few billion years. For these reasons, scientific revolutions in geology are based primarily on reconstructing events in theory rather than reproducing them in fact. Even so, great things may be discovered through a combination of observation and reason.

Leonardo da Vinci Shows How Geological Field Data Forced Him To Doubt the Occurrence of the Biblical Flood

Leonardo da Vinci (1452–1519) showed a spark of geological genius along with his other remarkable qualities. His childhood was spent at Vinci, just north of the Arno River some 27 kilometers (17 miles) west of Florence, in Tuscany, Italy. This area is far from the sea and quite unlikely to be subject to a flood like that described in Genesis. In his *Note-Books,* under the heading "Doubt," Leonardo considered whether or not the Biblical flood could have occurred as a natural phenomenon without divine intervention.

> Here a doubt arises, and that is whether the Flood which came at the time of Noah was universal or not, and this would seem not to have been the case for the reasons which will now be given. We have it in the Bible that the said Flood was caused by forty days and forty nights of continuous and universal rain, and that the rain rose ten cubits [15 feet] above the highest mountain in the world. But, consequently, if it had been the case that the rain was universal, it would have formed in itself a covering around our globe which is spherical in shape; and a sphere has every part of its circumference equally distant from its centre; and therefore, on the sphere of water finding itself in the aforesaid condition, it becomes impossible to move since water does not move of its own accord except to descend. How then did the waters of so great a Flood depart, if it proved that they had no power of motion? If it departed, how did it move, unless it went upwards? *At this point natural causes fail us, and therefore in order to resolve such a doubt we must needs* [*it is necessary*] *either* [*to*] *call in a miracle to our aid or else say that all this water was evaporated by the heat of the sun.* [Italics mine.][2]

Leonardo da Vinci has begun to see a crisis of contradiction. His word for it is *doubt.* In the next passage, entitled "Of the Flood and

Figure 2-2 *Leonardo da Vinci (1452–1519) saw that geological field data are valid testimony of events that occurred in prehistoric times, ". . . since things are far more ancient than letters. . . ."*[2] *("Photo," Science Museum, London)*

Marine Shells," we find him carrying on an argument with himself. It is a long discussion triggered by an observation of ". . . shells which are visible at the present time within the borders of Italy, far away from the sea and at great heights . . . ."[2]

> Should you say that the nature of these shells is to keep near the edge of the sea, and that as the sea rose in height, the shells left their former

> place and followed the rising waters up to their highest level: — to this I reply that the cockle is a creature incapable of more rapid movement than the snail out of water, or is even somewhat slower, since it does not swim, but makes a furrow in the sand, and supporting itself by means of the sides of this furrow, will travel between three and four braccia [arm lengths] in a day; and therefore, with such a motion as this could not have traveled from the Adriatic Sea as far as Monferrato in Lombardy, a distance of two hundred and fifty miles [400 kilometers] in forty days, — as he had said who kept a record of that time.[2]

We may glimpse the way that this crisis of contradiction was thrust on Leonardo by putting together two sets of thoughts. In the first one, we see Leonardo, the observer, reading the fossil record.

> [Shells] . . . one old, another young, one with an outer covering and another without, one broken and another whole, . . . one filled with sea sand and the fragments great and small of others inside the whole shells which stand gaping open; . . . with shells of other species fastened on to them, . . . found among the bones and teeth of fish.[2]

In the second, he acknowledges the impact of the field data.

> *Since things are far more ancient than letters,* it is not to be wondered at if in our days there exists no record of how the aforesaid seas extended over so many countries; and if moreover such record ever existed, the wars, the conflagrations, the deluges of the waters, the changes of speech and habits have destroyed every vestige of the past. *But sufficient for us is the testimony of things produced in the salt waters and now found again in the high mountains far from the sea.* [Italics mine.][2]

Unfortunately, Leonardo da Vinci's thoughts about the way geology must be studied in the field were not known to contemporary scholars. Nevertheless, a crisis of contradiction began to emerge as protogeologists discovered that rocks told more reliable tales of their own creation than those found in the Bible. After nearly three centuries James Hutton of Edinburgh developed a new, revolutionary paradigm in his book entitled *Theory of the Earth* (1788). Hutton's message was that rocks could be interpreted using knowledge of geologic processes that are still active on the earth today. A true science had been born by the turn of the nineteenth century.

Charles Darwin Discovers That Organic Evolution Must Be a Reality

The name Charles Darwin (1809–1882) brings to mind a graybeard of fifty years or more who astounded the world with his iconoclastic view of nature published in 1859 entitled, *On the Origin of Species.* It was a study of organic evolution and the way a process he called *natural selection* brings it about. *Organic evolution* means a systematic change in the forms of animals and plants with time. *Natural selection* is "a process by which less vigorous and less well adapted individuals tend to be eliminated from a population without leaving descendants to perpetuate an inferior stock."[3] *On the Origin of Species* did not spring full blown from Darwin's secluded study in Down House, in Kent, England. The new paradigm had a long history. We shall examine one small part of it, the anomaly that triggered his crisis of contradiction and forced him to see that organic evolution must be a reality.

His First Anomaly and the Recognition of a Crisis of Contradiction Darwin discovered anomalies at the age of twenty-two in a very odd way, through a hoax. He had been an indifferent student of both medicine and theology at the universities of Edinburgh and Oxford before becoming interested in natural science. In the summer of 1831, he was fortunate enough to be the field assistant to Adam Sedgwick (1785–1873), one of the greatest geologists of all time. They were mapping relationships of sedimentary beds (strata) in the central part of England, east of Wales. Darwin was told that the shell of a large tropical mollusk, *Voluta,* had been found in a gravel pit nearby. Darwin imagined that Sedgwick would be delighted at ". . . so wonderful a fact as a tropical shell being found near the surface in England."[4] On the contrary, Sedgwick recognized it as a hoax rather than a natural anomaly, saying ". . . it would be the greatest misfortune to geology as it would overthrow all we know about the surficial deposits of the Midland counties."[4] At this time, Sedgwick did not know that gravels were glacial deposits, but he did know that nothing like a tropical marine fauna had ever been found in them. This reasoning may have had a great deal to do with Darwin's future career. Apparently for the first time, Darwin realized that ". . . science consists in grouping facts so that general laws or conclusions may be drawn from them."[4] *On the Origin of Species* and his other books are truly massive collections of facts coupled with the general conclusions he drew from them.

New Horizons The year 1831 was a great one for Charles Darwin. In August, he received an invitation to sail on H.M.S. *Beagle* as "naturalist without pay" for a five-year, forty-thousand mile voyage of discovery. His quarters were cramped, with ". . . just room to turn around and that is all."[4] One part of his library, which did open almost unlimited horizons, was Charles Lyell's (1797–1875) revolutionary new paradigm, *Principles of Geology.* This three-volume set was the first truly modern model of geology. Lyell possessed an unmatched gift for reading the language of geology as it is written in the field. He saw it as a language expressed in minerals, fossils, rocks, structures, land forms, and, most importantly, in the processes that produce them all.

Lyell's books showed Darwin how to draw least-astonishment conclusions about the history revealed in the many rocks he encountered along the way.

Charles Lyell made two other important contributions to Darwin's work during the *Beagle*'s voyage. One was to draw his attention to the possibility of organic evolution. We see the evidence in some of the topic headings of Chapter 9, Volume I of the *Principles*. Darwin must have read it during the long weeks at sea outbound from England. These headings are still startling to scholars of evolution.

> Theory of progressive development of organic life considered . . . Evidence in its support wholly inconclusive . . . Vertebrated animals in the oldest strata . . . Differences between the organic remains of successive formations . . . Remarks on the comparatively modern origin of the

Figure 2-3 *Sir Charles Lyell (1797–1875) was the most famous and, perhaps, most able British geologist of the nineteenth century.* Principles of Geology, *produced from his own field observations, was the first modern textbook in this science. (Courtesy of the Library of Congress)*

human race . . . The popular doctrine of successive development not confirmed by the admission that man is of modern origin . . . In what manner the change in the system caused by the introduction of man affects the assumption of uniformity of the past and future course of physical events . . .[5]

Lyell's other principal contribution to Darwin's work was to emphasize the vast geological time available for the slow modification of species. We know how important his contribution was, for Darwin clearly acknowledged the debt when he wrote his own book nearly three decades later. "He who can read Sir Charles Lyell's grand work on *Principles of Geology* . . . yet does not admit how incomprehensively vast have been the past periods of time, *may at once close this volume.*" [Italics mine.][6]

Darwin's destiny, at the moment of sailing, was predicated on these four factors: 1) he had the tools with which to work; 2) he knew he had the tools; 3) he had the whole world before him as his laboratory; 4) he knew that, too.

Crisis of Contradiction: Finches of the Galapagos Islands
Perhaps the most critical moments for Darwin came in the Galapagos Islands during three weeks in September and October of 1835. Darwin seems to have been alert to the significance of anomalous data as the key to a new view of the world. His experience with *Voluta* may have taught him to search for unexpected variations in otherwise normal patterns. Nothing would seem to be more commonplace than to find ground finches among the birds inhabiting the islands that lie on the equator, some 650 miles west of Ecuador in the Pacific Ocean. Darwin noted that finches on different islands were distinguished by the shapes of their bills. He did not know what to make of the data but, sensing an anomalous situation, placed this note in his journal: "I have stated, that in the thirteen species of ground-finches, a nearly perfect gradation may be traced, from a beak extraordinarily thick, to one so fine that it may be compared to that of a warbler."[4] In those days, naturalists were unsure of the meanings of variations within species. They still thought of species in rather rigid terms.

The meaning of Darwin's observation lay fallow for some time, even after his return to England. By 1845, a full decade later, another note appears in the second edition of the *Journal of Researches of the Voyage of the* Beagle: "Seeing this gradation and diversity of structure in one small, intimately related group of birds, one might really fancy that from an original paucity of birds in this archipelago, one species has been taken and modified for different ends."[4]

Figure 2-4 *This is Darwin's illustration showing the range of shapes of the bird bills found on the different finches of Galapagos Islands.*[7] *Imagine such simple data having such profound consequences as to trigger a scientific revolution.*

At this point, Darwin had begun to respond to the crisis of contradiction brought on by recognizing remarkable variations within the same species of birds. He could no longer accept the old idea that every member of a species is exactly like every other member. This left him no alternative but to consider modification of the species to account for the variations found on each of the different islands in the Galapagos group. That first intellectual step was more simple than the next one. Darwin knew that if variation occurred, there must be a mechanism to cause it. Many years were to pass before he solved that problem.[8]

Natural Selection, a Mechanism That Changes the Character of Populations Charles Darwin eventually developed a new, revolutionary view of nature in which entire populations of animals and plants are changed by a mechanism he called *natural selection.* This is a profoundly simple idea. Every animal and plant lives in a natural environment that both supports it and exerts pressures on it. Rain, drought, heat, cold, predators, disease, the amount of available food, and accidents are typical pressures that may shorten the natural life span of any individual. Some individuals in a given population may be fit enough to survive every pressure put on them. Other individuals may not be fit enough and may die off before they pass on their weak characteristics to the general population. Only the surviving part of the original population is important to future generations.

Imagine that survival depends upon longer fangs, better claws, or superior brain power. Individuals who lack these qualities tend to die out early before passing on these weak characteristics. The result is a shift in the qualities possessed by the next generation. Evolution is a pragmatic, cold-blooded procedure. There is no substitute for success! That is exactly what natural selection means. Individuals are selected through natural stresses to give the next generation the qualities that make it fit enough to survive.

Organic evolution is now known to be an enormously complicated process. Its general acceptance by the modern scientific community depends on the mutual support of a great mass of interlocking facts. In Darwin's time, practically nothing was known about the science of genetics and heredity. Natural selection may have seemed mystical in 1859. That is no longer true. The process is now known to be a genetic one rather than some mechanical limitation of the survival of future generations.

Much of the resistance to the evolutionary viewpoint has come from people who object to the inevitable conclusion that human beings are animals produced by, and caught up in, the processes of change. The Darwinian revolution, like the Copernican revolution before it, has managed to move the human race a little farther from the center of the universe.

Alfred Wegener Challenges the Concept of the Permanence of Continents and Ocean Basins with Another Idea: Continental Drift

Our third illustration of the part played by anomalies in scientific revolutions, the idea of continental drift, is especially interesting because the intellectual event is still taking place. Like the evolution revolution, the anomalous data were accumulated for nearly a century before the crisis stage was reached. In this case, the new view of nature accounts for the assembly and disassembly of parts of the crust of the earth (called plates). These plates are very large; for example, the one containing North America extends from the center of the Atlantic Ocean westward to the rim of the Pacific Ocean. The plates float on very large flowing currents that move slowly within the body of the earth and carry the continents along with them. Linear mountain ranges bordering continental rims have been produced by the interaction of rock masses along plate boundaries. The full revolutionary story of this mountain building machinery has not yet been deciphered, although the major portion of the crisis of contradiction has been passed. The next phase of the revolution will consist of the discovery of more anomalies, contradictions, and subcrises. Many decades of research are required before the full revolution can be completed.

The new paradigm has been called by such various names as *continental displacement, sea-floor spreading, plate tectonics, and mantle convection.* We are not concerned with the technical aspects of the present frontiers of knowledge as much as we are with the original breakthrough that triggered the crisis of contradiction. Wegener, like Darwin, recognized the meaning of an anomaly. Up until Wegener's time, the ruling paradigm was a belief that continents and ocean basins were permanently fixed in their present places on the face of the earth.

Wegener's View of the Crisis of Contradiction Alfred Wegener (1880–1930), a German meteorologist and geophysicist, triggered the revolution. His first awareness of the anomaly came when he recognized that the shapes of the continental outlines on both sides of the Atlantic Ocean appeared to be nearly identical. This suggested that he was looking at broken and separated parts of once-connected land masses. Wegener was no parlor-armchair revolutionary. His premature death in 1930 occurred in Greenland where he was attempting to find field evidence to prove that Europe and North America had once been joined. He felt that the parts should fit together as well as pieces of a torn newspaper.

Figure 2-5 *Alfred Wegener (1880–1930) was a German geophysicist who recognized that continental drift was his best answer to a crisis of contradiction. (Reproduced by kind permission of Blackie and Sons Ltd., Glascow and London)*

Figure 2-6 is Wegener's model of the breakup of a single great continental unit. This map was published in 1924 as part of his famous book, *The Origin of Continents and Oceans.* The whole idea of continental drift might have been forgotten if the only anomaly were the fit of the borders on each side of the Atlantic Ocean.

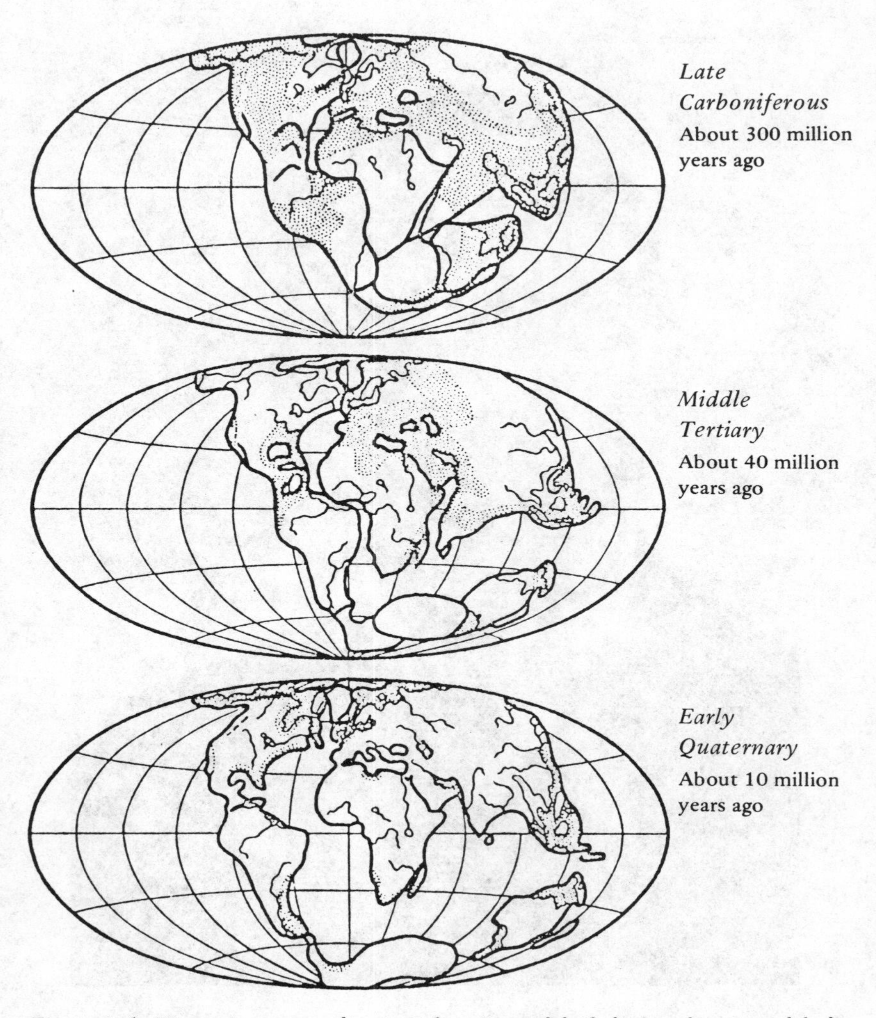

Figure 2-6 *Wegener's conceptualization of continental drift during the course of the last 300 million years. Stippled areas on these maps mark positions of shallow seas that partially covered the continental blocks from time to time. (From Alfred Wegener,* The Origin of Continents and Oceans, *1924. Translated from the third German edition by J. G. A. Skerl. Dover Publications, Inc., 212 pp. Fig. 1, p. 6, has been used as a base for this illustration by permission of the copyright holder.)*[9]

However, there was another strange fact that had been puzzling geologists for decades before Alfred Wegener suggested an explanation for it.

Thick sedimentary deposits about 250 million years old and of

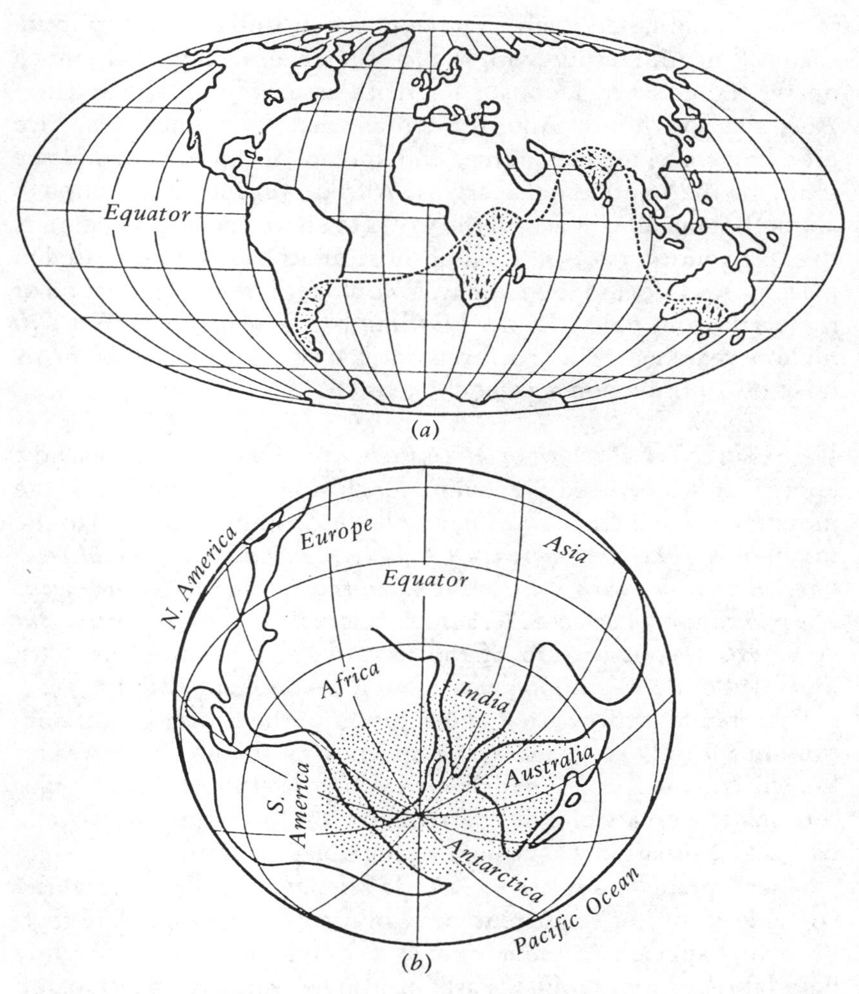

Figure 2-7 *Distribution of 250 million year old glacial deposits, the anomaly that triggered a scientific revolution. A, Areas covered by continental glaciers at that time. The arrows show the direction of ice flow. B, Areas glaciated 250 million years ago and reassembled at the South Pole, as interpreted by Wegener in 1924. (After Arthur Holmes,* Principles of Physical Geology, *2nd ed., 1965. Courtesy T. D. Nelson and Sons, Ltd., Sunbury-on-Thames and Ronald Press, N.Y.)*[10]

unmistakable glacial origin were known to rest on ice-scarred and ice-grooved pavements of older rocks in parts of South America, South Africa, India, and Australia. The distribution of glacial ice and direction of flow in these areas is shown in Figure 2-7(a). Here is an astounding anomaly! The deposits in India extend to more than 30° north latitude with the ice center far to the south, much nearer the equator. Deposits in Africa actually cross the equator. Australia and South America contain glacial sediments and ice grooving that indicate sources and spreading centers beyond the limits of the continental margins. Alfred Wegener knew enough about the ancient climates of the world to appreciate the meaning of this distribution pattern. If all of those areas had been glaciated in place, just as we find them today, it would require total glaciation of the earth from pole to pole. Nothing like that had occurred 250 million years ago. The rock record is very clear on this point. A crisis of contradiction was seen to exist.

Wegener's New Paradigm of Continental Drift We can now appreciate why Alfred Wegener's conclusion was inevitable. If the mountains won't come to Mohammed, Mohammed must go to the mountains. *The continents must have been moved from original positions close to the South Pole during glaciation to their present positions after glaciation was over.* Wegener called this motion *continental drift.* His reconstruction of the way the South Pole must have looked about 250 million years ago is given in Figure 2-7(b).

Continental drift was a grand solution to the crisis of contradiction, but it only served to bring up a new problem. There was no known force capable of driving continents thousands of miles through the crusts of solid basalt known to underlie every ocean basin. The situation was again unacceptable.

Unacceptable is a relative word. In science, it really means unacceptable according to current views of nature. Every standard textbook first appears on the market as a compilation of the currently held fabric of a discipline. Parts of it may even become outdated before publication through the surprising appearance of an important research report. New discoveries are cumulative and force adjustments in the growing models of nature. Old texts become histories of the record of outmoded thoughts of former days. A study of a series of successive textbooks is often very rewarding, for they show exactly how scientific progress is made.

We shall look at an ideal example, a set of texts in historical geology spanning the five decades during which the Wegnerian revolution occurred. Four different authors collaborated on successive editions. Their longevities and overlapping professional lives led to a set of books containing an unbroken chain of thought. The first book, published in 1915, was contemporary with the first German edition of Wegener's book. Naturally, there is no mention of continental drift in it. The last revision, in 1969, was written to catch up with the revolution.

The Status of the Old Paradigm Circa 1920 Charles Schuchert was a most respected professor of geology at Yale when he published his marvelous compendium *Historical Geology* (1915). The basic paradigm, which Wegener so successfully challenged, is found under the heading, "Permanency of Continents and Ocean Basin." The standard beliefs of the time are set out in the following excerpts:

> . . . most geologists hold with Dana (another great Yale professor, 1813–1895) that the oceanic basins and the continents have in the main, although not in detail, been permanent features of the lithosphere at least since the close of Proterozoic time (about 600 million years ago). There is likewise much agreement among geologists in the belief that the oceanic basins are sinking areas, also spoken of as the "negative areas" of the lithosphere, because the sum of their crustal movements is downward; and in general it appears that the oceanic basins have gradually attained not only greater depth but somewhat enlarged area as well. On the other hand the continents are the rising masses of the lithosphere in relation to sea level, and for this reason are also called the "positive areas," because the sum of their movements is upward . . . In general, however, it may be said that the present oceans and continents have been more or less permanent features, and that they have been where they are now, with moderate changes in their outlines, since their origin in Proterozoic time or earlier.[11]

In those days, geologists were unable to identify all the different kinds of deep-sea deposits encountered in the field. Professor Schuchert imagined that deep-sea deposits were much more rarely represented within the continental sediments than they really are.

He considered their rarity to be substantial proof of the permanence of continents and ocean basins. At that time, his additional proof appeared to be quite sound.

> It is now established that the lithosphere is denser and therefore heavier beneath the ocean basins than under the lands, making it impossible for them to have interchanged their positions without destroying the equilibrium of the outer shell. . . .[11]

Geologists refer to this balance as an *isostatic* condition. The word is from the Greek *isos,* meaning "equal," and *statikos,* meaning "causing to stand." An isostatic condition is one in which the great land forms stand in nearly perfect balance, with the same amount of weight pressing down on the interior of the earth beneath each large unit of area.

Lurking Crisis: Continental Fragmentation The crisis of contradiction is often expressed so subtly in nature that scientists fail to notice it. After Schuchert had made a strong case for the permanency of continents and ocean basins, he became enmeshed in the inconsistencies of his own model. So he created another one to go with it. He presents the second model with an element of creeping seduction, almost as if he were forced to come to terms with the field data a little at a time. He agreed that the floor of the Mozambique Channel between Africa and Madagascar must have sunk beneath the sea after the evolutionary development of the familiar African animals. This sinking explained their presence on Madagascar. The sinking was not too hard to conceive, for the intermediate Comoro islands served as evidence that Africa and Madagascar had indeed formerly been joined by dry land. However, once Schuchert had gone this far, he was forced to consider the Seychelles, a group of islands about 1,100 kilometers (700 miles) northeast of Madagascar. They, too, are made of continental rock types rather than rock types common to ocean basins.

At this point in his thinking, Schuchert toyed with the idea of a large land mass, *Lemuria,* that must have foundered beneath the sea in comparatively recent times. Here is the story in his own words:

> Examples of Continental Fragmenting. — As an example may be cited Madagascar, . . . No naturalist doubts its former connection with Africa, because of their similar animals, and yet the channel of Mozambique which now separates it from the mainland is from 240 to 600 miles [386 to 965 kilometers] wide . . . To the northeast of Madagascar lie many small islands, the Seychelles, and to the northwest the Comoro group, all of which are also held to have been parts of Africa and Madagascar. Not only this but many biologists and geologists hold that all of these lands are but parts of the comparatively recent land Lemuria . . .[11]

The next steps in his thinking were truly drastic deviations from the paradigm of the permanence of continents and ocean basins. He joined Africa and South America by replacing the Southern Atlantic Ocean with a land bridge he called *Gondwana Land,* and the northern Atlantic Ocean by a land bridge he called *Eria* or *Holarctica.*

> Besides the facts given above, there is much other evidence of a geologic, paleontologic, and zoologic character relating to the distribution of plants and animals since the Paleozoic (about 230 million years ago) tending to show that Brazil was once widely connected with northwestern Africa across what is now the deep Atlantic Ocean. This lost continent is the "Gondwana Land" [from a district of the same name in India] of Neumayr (1883) and Suess (1885), and of the zoogeographers, a vast transverse land stretching from the northern half of South America, across the Atlantic to Africa and thence across the Indian Ocean to peninsular India, including Lemuria. It was in existence throughout the Paleozoic [geological era extending from about 600 million years ago to 230 million years ago], but the Atlantic bridge and Lemuria sunk into the oceans during the Mesozoic [geological era extending from about 230 million years ago to 65 million years ago]. Gondwana, when complete, was comparable to another transverse land of the north, "Eria" or Holarctica, which existed when North America

was continuous with Greenland and Eurasia across Iceland to the British Isles.[11]

Later in the book, Schuchert considered the difficulties of explaining Permian continental glaciation so near the equator on both the north and south sides, but chiefly in the southern hemisphere. This was the same anomaly that bothered Wegener, forcing him to consider the possibility of continental drift. Schuchert was not ready to go that far in 1915, but, by the time he had written 356 pages of his book, he did recognize the crisis of contradiction. Here he states it clearly:

> Belief in the existence of Gondwana is widespread among European geologists, but many American workers do not yet believe in it, mainly because they hold strongly to the theory of the permanence of the oceanic basins and continents. Without this continent, on the other hand, paleontologists cannot explain the known distribution of Permian landlife and, further, its presence is equally necessary for the interpretation of the peculiar distribution of marine faunas beginning certainly with the Devonian [405 to 350 million years ago] and ending in the Cretaceous [135 to 65 million years ago].[11]

In 1920, Schuchert pointed out critical happenings and the dates on which they occurred. His evidence was sound. The global aspects of faunal distribution in time, assembly of plates into supercontinents, and disassembly of them again into the modern fragments we call by familiar names would become parts of the revolutionary new paradigm of the 1960s called global tectonics. Unfortunately, Schuchert did not live to see his timidly expressed vision vindicated.

The Status of the Old Paradigm Circa 1941 The 1941 edition of *Historical Geology* was co-authored by another great Yale professor, Carl O. Dunbar. The prologue in this edition shows the authors have not changed greatly their basic ideas about permanent continents and ocean basins.

> The oceans have always been where we see them now . . . The floods (seas that spread across the continents) are as often withdrawn by

recurrent deepening of the oceans, but there never has been general interchange in position between the continental masses and the basins of the oceans.[12]

Lurking Crisis: Permian Glaciation Within 20° of the Equator Is Best Explained by Continental Drift Later in the same chapter, the authors reflect a softer attitude toward continental permanence as they face the data that helped convince Wegener.

> The most remarkable feature of the Permian glaciation is the *distribution* of the ice sheets. They were chiefly in the southern land masses and in regions which now lie within 20° to 35° of the equator. This circumstance more than any other lent attractiveness to the belief in "continental drift." If the southern continents were united to Antarctica until after Permian time, the glaciation may not have spread into low latitudes. A later "drift" of these continents toward the north would account, far more easiy than any other means yet postulated, for the present distribution of the glacial deposits. But this premise itself is still in the realm of speculation![12]

The Status of the Old Paradigm Circa 1960 Two decades later, the crisis of contradiction was about to erupt into a full-blown revolution. By this time, Schuchert had died and Dunbar was carrying on alone.[13] The rigid ideas about the permanence of continents and ocean basins are missing in the third edition, but the comments on Permian glaciation remain unchanged. The author does not straddle the paradox and he does not accept drift; he abstains. In the next few years, new geophysical evidence overwhelmed the last resistance, but in 1960 the decaying paradigms still held firm. A fourth printing of the 1949 text was made in February of 1962.

Revolution Circa 1969: Wegener Wins The 1969 edition, coauthored with a third Yale professor, Karl M. Waage, caught up with the times.[14] A new chapter entitled "The Restless Crust" was added to replace the abandoned ideas about the permanence of continents and ocean basins. The authors chose a verse from Tennyson to

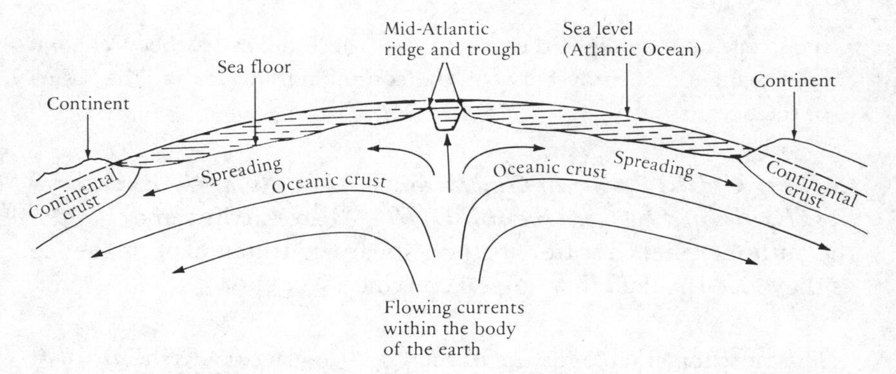

Figure 2-8 *A few scientists of genius saw the new view of nature as a dynamic model of the earth. In this paradigm, the rocky interior contains flowing currents that seem to rise from depths of about 1,000 kilometers (625 miles), split, and flow off around the earth in two directions from the positions of the midoceanic ridges. New oceanic crust forms and fills in the gap at the split. Continents are carried along with the flow.*

express the spirit of change that followed the Wegenerian revolution.

> There rolls the deep where grew the tree,
> O Earth what changes hast thou seen!
> There where the long street roars hath been
> The stillness of the Central sea.[15]

There are many more parts to the story. Geophysical evidence began to appear as early as 1735. Enough information had been accumulated by 1855 to prove that the Himalaya Mountains have hidden roots and that this might be true of other linear mountain ranges as well. No one suspected at the time that this great ridgepole on the roof of the world was formed when the Indian subcontinent drifted northward across the equator and collided with southern Asia. The Permian glacial deposits simply rode along on the back of India. Proof was established in the late 1950s with the discovery of the structure of the basaltic, dark, fine-grained, iron-rich, igneous rocks of the oceanic crust. A few scientists of genius saw the new view of nature as a dynamic model of the earth. In this paradigm, the rocky interior contains flowing currents that seem to rise from depths of about 1,000 kilometers (625 miles), split, and flow off around the earth in two directions, centered on the midoceanic ridges. New portions of oceanic basaltic crust form and fill the gap where the split takes place. Figure 2-8 illustrates how the continents are carried along on top of the moving subcrustal currents. Alfred Wegener had won!

2-4 Joining the Scientific Fraternity

Student learning is not so different from that of professional scientists. Anomalous facts and the crisis of contradiction drive any thinker toward understanding. Students receive a great deal of information that does not fit into their world views. The data may be handled in two ways. If you learn them by rote without appreciating their meaning, you remain outside the scientific community. But when cognitive learning becomes habitual, you have joined the community. Figure 2-9 describes learning by comparing the students' fabric of scientific data with the scientists'.

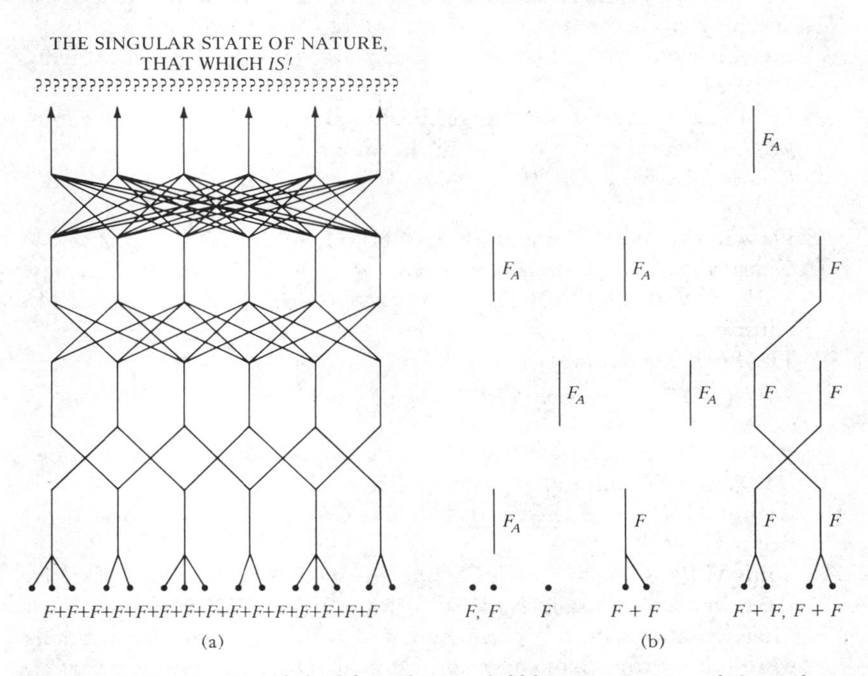

Figure 2-9 *Comparison of the fabric of science held by scientists (a) and the students' view of it (b). Students see isolated, unexplained facts,* F_A, *as anomalies. (From John W. Harrington,* To See a World, *St. Louis: The C. V. Mosby Company, 1973)*

Annotated References

1. Thomas S. Kuhn, *The Structure of Scientific Revolution,* 2nd ed., International Encyclopedia of Unified Science (Chicago: University of Chicago Press, 1970), II.
The definition of normal science is quoted from p. 10. His brief definition of the paradigm is taken from the preface, p. viii. The shared characteristics of paradigms are found on p. 10. The quotation about discovery and the awareness of the anomaly is found on pp. 52–53. The final quotation dealing with discovery and the scientific revolution is from pp. 89–90.
2. Leonardo da Vinci, *Leonardo da Vinci's Note-Books,* ed. Edward MacCurdy (New York: Empire State Book Company, 1923).
3. *American Geological Institute Dictionary of Geological Terms* (Garden City, N.Y.: Doubleday, 1962).
4. Sir Gavin de Beer, *Charles Darwin* (Garden City, N.Y.: Doubleday, 1965).
The quoted passages begin on p. 30. All direct reference to Darwin and his writings are taken from this source, pp. 30, 34, 82, and 132.
5. Sir Charles Lyell, *Principles of Geology* (London: John Murray, 1830), I.
Lyell's awareness of the changes in animals and plants through geological time is taken from the topic headings for chapter 9, p. 144.
6. Charles Darwin, *On the Origin of Species* (London: John Murray, 1859).
Darwin's acknowledgment of his debt to Lyell is found on p. 282.
7. Charles Darwin, *Journal of Researches of the Voyage of the* Beagle (New York: P. F. Collier, 1902), pp. 587. (A reprint of the famous 1845 edition.)
The finches are found on p. 429.
8. Ernst Mayr, "The Nature of the Darwinian Revolution," *Science,* 176 (1972), 981–989.
9. Alfred L. Wegener, *The Origin of Continents and Oceans,* 3rd ed., trans. J. F. A. Skerl (London: Methuen, 1924).
10. Arthur Holmes, *Principles of Physical Geology,* 2nd ed. (New York: Ronald, 1965).
11. Louis V. Pirsson and Charles Schuchert, *Historical Geology, A Textbook of Geology,* 2nd rev. ed. (New York: John Wiley, 1920).
The quotations on the permanency of continents and ocean basins were taken from Schuchert and Dunbar, *Historical Geology,* part II, pp. 463–466. Comments on the glacial climate of the Permian were

quoted from pp. 758–760. Finally, Professor Schuchert's summary of the paradox of continental glacial deposits in India and on both sides of the equator is found on p. 761.

12. Charles Schuchert and Carl O. Dunbar, *Historical Geology: A Textbook of Geology,* 4th ed. (New York: John Wiley, 1941).
The prologue was quoted from p. 2 of *Historical Geology.* The section on Permian continental glaciation was taken from p. 291.
13. Carl O. Dunbar, *Historical Geology,* 2nd ed. (New York: John Wiley, 1949).
14. Carl O. Dunbar and Karl M. Waage, *Historical Geology,* 3rd ed. (New York: John Wiley, 1969).
The reference and verse were taken from p. 69ff.
15. Alfred, Lord Tennyson, *In Memoriam,* ed. William J. Rolfe (Boston: Houghton Mifflin, 1895), p. 206.
"There rolls the deep . . ." begins stanza 123. The other verses in the stanza are also familiar to geologists, and indicate the degree to which geology pervaded the world of letters in the nineteenth century.

THREE

CLASSIFICATION: THE SEARCH FOR SYSTEM AND ORDER

If you see a stick in the road you may say,
"That looks like a snake."
But if you see a snake in the road you never say,
"That looks like a stick."

Old South Carolina Low-Country Saying

3-1 Nature's Bewildering Display of Facts

Nature presents us with a bewildering display of facts that can be understood best by first recognizing their beautiful system and order. Scientific classifications are hard-won models of these ordered systems. We will learn to appreciate them, heart and soul, by examining how the pioneers of science discovered a few interesting examples. Students will recognize similarities between the struggling efforts of famous scientists and their own agonies of learning. These similarities are not superficial ones. In both cases, we are concerned with the birth of knowledge.

The task of classifying nature is very old. Our immediate ancestors evolved several million years ago. Much classifying was done long before there was anything we would recognize as a genuinely human civilization. Consider, for example, the host of geniuses who invented spoken languages thousands of years before the idea of formalized science was ever conceived. They made up words to stand for groups of things, thereby simplifying the number of items to be named and understood.

The word *food* designates all things that can be eaten to support life. Think of all the items included and excluded by this one classifying word. Chocolate-covered ants and raisins are food; chewing gum is not. Our modern English word may be traced back through its Anglo-Saxon root, *foda*, Old Norse, *faeve*, Teutonic *fa* or *fo*, to Old Aryan *pat*. This trail leads far back in time across northern Europe, southeast to the Iranian plateau and the misty origins of the human race on the vast continents of Africa and Asia. The idea of food is certainly older than our species. Even the sound of the word has a throaty, animal quality as air is expelled from the chest and out over the lips . . . "food". Good classification systems are practical. They are tied to the nature of nature and, in this way, are timeless and independent of shifting human cultures.

This does not mean that our perception of nature and classified views of its system and order must remain fixed. The idea of a *star* in the sky is a good example. This modern English word also has an ancient Mideastern root. There are billions of stars, but long ago someone recognized a way to communicate meaning by assigning a single word to stand for any member of the group. In time, that word became an oversimplification. Subdivisions were necessary to account for the unblinking planets that wandered across the patterns of fixed stars. What a shock it must have been to recognize for the first time that the sun is a star, too! It must have been frightening to realize that the sun appears to be larger than the rest of the stars simply because it is so close to us. The invention of powerful telescopes led to the recognition of billions of galaxies, each made up of billions of individual stars. More recently, astronomers have found that the universe contains cepheid variables (regularly pulsing stars), quasars (quasistellar objects), and black holes, from which no revealing radiation may escape. Classification systems grow as science grows. That is obvious. The way they help science grow is much less obvious.

Classifications do not appear out of thin air. They are artifacts that, at the time they are conceived, represent the limit of imagination tempered with scientific judgment. This is also exactly how we learn. Our imaginations sprawl into the unknown and select what can be understood through tempered judgment. Follow the step-by-step histories of these next two examples to see how the pattern unfolds. Exploration of the history of science is an excellent way to learn to think as professional scientists think.

3-2 A Discussion of Two Critically Important Classification Systems

"I can't hear you until I know where you're coming from," was a popular slang expression among students of the 1970s. The idea is a good one. It is much easier to understand something if we know its background and can see it in its context. Scientific classification systems have been pieced together bit by bit from partially recognized information. Studying them is much more fun if you see them as part of an open-ended intellectual struggle rather than as highly polished, finished products to be memorized. The following illustrations demonstrate some of the blundering approaches that have plagued scientists for centuries. Many pioneers never knew the success to which their work would eventually lead.

The Biological Classification of Animals and Plants

Children become aware of living things in a more or less random way. Different flowers, trees, dogs, pigs, cats, and cows are seen as individuals deserving this name or that one because they seem to possess similarities with some imaginary standard prototypes. A cow is a cow because it possesses unmistakable cowishness. Nothing could be simpler or more certain than that — at first glance. This is the world filled with cats and dogs, bats and rats that Swedish biologist Karl von Linne (1707–1778) inherited.

Linne spent much of his professional life in a devoted effort to bring order to the chaos found in the traditional methods of biological classification. Writing in Latin as Carolus Linnaeus, he published twelve major works on systematic classifications of thousands of plants and animals. His most famous book was *Systema Naturae* (1735). After many editions, this work eventually contained descriptive classifications of 4,235 different animals. *Genera Plantarum* (1737), *Classes Plantarum* (1738), and *Species Plantarum* (1753) were equally important. About 8,000 plants were classified in these volumes.

Linnaeus (as we shall now call him) made two outstanding contributions to the development of a classification system. He standardized the use of two names at the level of genus and species for every living form. He also popularized a standard set of classification categories. The modern counterpart of his system is illustrated in Figure 3-2 with a display of relationships that exist between house cats, lions, tigers, and the extinct saber-toothed tiger. Linnaeus's system is easy to understand.

These are all animals. Therefore, they belong in the animal kingdom rather than the plant kingdom. Each has a main nerve cord running down the back, so they all belong to the phylum *Chordata.* They all have bones fashioned into a skeletal system with vertebrae surrounding the nerve cord. This places them all in the subphylum *Vertebrata*. They are all four-footed. Therefore, they belong in the superclass *Tetrapoda*. The females have mammary glands to supply milk to nurse the young. That trait puts them in the class *Mammalia*. All are meat eaters; therefore, they belong to the order

Figure 3-1 *Swedish biologist Karl von Linne (1707–1778) invented the two-name classification system for animals and plants that is still used in modified form today. His name is more familiar in the Latinized version, Carolus Linnaeus. (Courtesy of the Library of Congress)*

Carnivora. They look like cats. This trait is unmistakable and includes them in the family *Felidae*. Up to this point, all traits have been grouped by similarities rather than differences. Beyond this point, differences as well as similarities distinguish one subfamily from another.

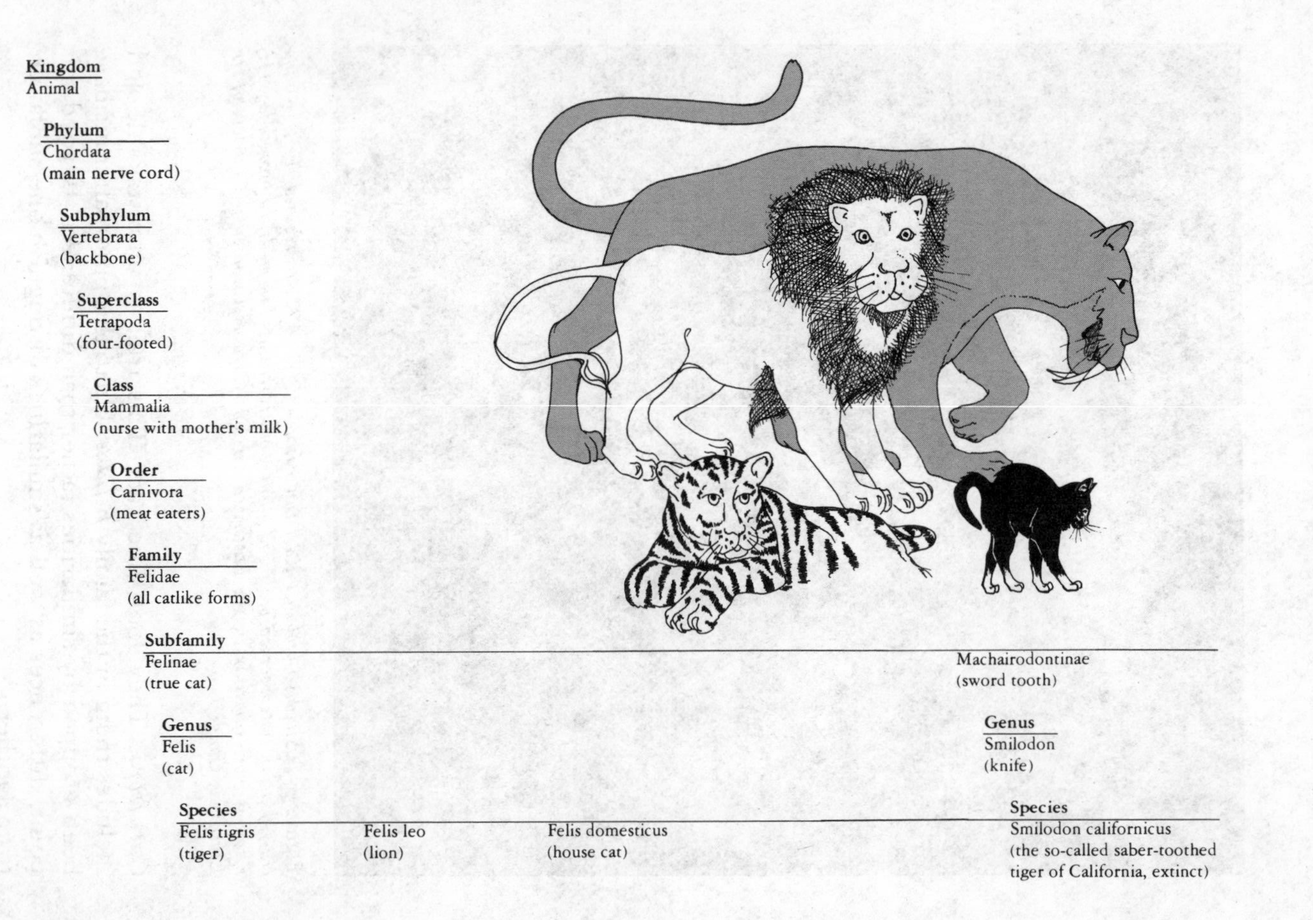
Kingdom
Animal
Phylum
Chordata
(main nerve cord)
Subphylum
Vertebrata
(backbone)
Superclass
Tetrapoda
(four-footed)
Class
Mammalia
(nurse with mother's milk)
Order
Carnivora
(meat eaters)
Family
Felidae
(all catlike forms)
Subfamily
Felinae
(true cat)
Machairodontinae
(sword tooth)
Genus
Felis
(cat)
Genus
Smilodon
(knife)
Species
Felis tigris
(tiger)
Felis leo
(lion)
Felis domesticus
(house cat)
Species
Smilodon californicus
(the so-called saber-toothed
tiger of California, extinct)

Since house cats, lions, and tigers are anatomically similar, they are placed together in the subfamily *Felinae*. However, the saber-toothed tiger was quite different. It had monstrous canine teeth extending downward from the upper jaw. This structure is shared by many other members of the family *Felidae* and justified establishing the subfamily *Machairodontinae*. The root words, *machairodus* and *odons*, are Greek for *sword* and *tooth*.

Linnaeus's work is best known for the idea that each plant and animal must have two names: the first indicating the genus to which it belongs, the second the species. The tiger, lion, and house cat all belong in the genus *Felis*, but they represent different species. Therefore, their names are *Felis tigris* (cat tiger), *Felis leo* (cat lion), and *Felis domesticus* (cat house). The saber-toothed tiger was not really a tiger at all. It did not belong to the same subfamily as the true cats and was part of a different genus, *Smilodon* ("knife"). The proper name of the saber-toothed tiger found in the La Brea tar pits near Los Angeles, is *Smilodon californicus* ("knife of California").

The word *genus* is from the Latin and means "race, group, or kind." Biologically, it means an ancestral group of plants or animals from which uniquely different *species* are descended. The word *species* is much harder to define, for its meaning has shifted at least twice since it was first used. This is the most interesting part of the story. It shows outstanding scientists as confused human beings, struggling to learn their trade while a scientific crisis of contradiction goes on around them.

In the early part of the eighteenth century when Linnaeus began his work, the concept of a species was very rigid. It was based on the idea of a model specimen, or archetype, as a standard. This idea conformed to the religious dogma of the day. Every plant and animal was assumed to have been created in its final perfection. Species were not believed to change. Their characteristics were supposed to have been fixed for all time by God during the fateful six days of Genesis. Here is the way Linnaeus stated the idea in his book *Classes Plantarum*: *"There are as many species as there were*

Figure 3-2 (facing page) THE BIOLOGICAL CLASSIFICATION OF SOME MODERN CATS AND A SABER-TOOTHED TIGER

originally created diverse forms." [Italics mine.][1] The only exceptions to this rule were domesticated animals, with which people had tampered. Many systematists who tried to follow Linnaeus's thinking were puzzled by obvious variations among members of every species. Too many of these members were far too different from any archetype to justify the assumption that Linnaeus had made about species. Variation within species proved to be the anomaly that helped trigger a crisis of contradiction for Darwin and led to a revolution in the meaning of the word *species.*

It is easy to see how Linnaeus could assume that a species may be represented by a single archetype serving as a common denominator for all the variations. Our hands, feet, heads, legs, arms and vital organs do resemble one another. We recognize each other as members of the same species without confusion. Our method is the one children use, recognition based on familiarity. The difficulty arises when we must distinguish between other living forms with somewhat similar characteristics. Linnaeus introduced the concept of *genus* to acknowledge similarity without claiming identity.

That was fine as far as it went, but what of the variation among individuals within a species? Variation does exist and contradicts any attempt to establish a rigidly defined, standard individual (or archetype).

Variation within a species would be hard for any scientist, no matter how indoctrinated, to ignore. It is simply too obvious to escape notice. All modern humans are classified in the genus, *Homo* and the species *sapiens*; yet no two people are alike. Consider the variation among the audience in a crowded theater. Some are fat; some are thin. Some are tall; some are short. Some have large, heavy bone structures; others have small, fragile bones. Hair color (original) varies from individual to individual. Skin tints vary. Who among us could claim to be the human archetype?

An outstanding French biologist, Georges Louis Leclerc, Comte de Buffon (1707–1788), made an issue of this flaw in Linnaeus's ideas of classification. He felt that Linnaeus imposed an arbitrary order on living forms that does not exist in nature. Buffon wrote:

> This manner of thinking has made us imagine an infinity of false relationships between natural beings . . . It is to impose on the reality of the Creator's work the abstractions of our mind.

Figure 3-3 *French biologist Georges Louis Leclerc, Comte de Buffon (1707–1788) challenged the intellectual foundations on which Linnaeus had built his classifications of animals and plants. (Courtesy of the Library of Congress)*

> The more one increases the number of divisions in natural things, the closer one will approach the truth, since there actually exists in nature only individuals, and the Genera, Orders and Classes exist only in our imagination . . .
>
> One can descend by almost insensible degrees from the most perfect creature (man) to the most disorganized matter . . . It will be seen that these imperceptible gradations are the great work of Nature; one will find such gradations not only in size and form, but also in the motions, the generation and the succession of each species . . . It will be clearly perceived that it is impossible to give a general classification, a perfect systematic arrangement, not only for Natural History as a whole, but even for a single one of its branches.[2]

Buffon wrote these statements of frustration in 1749 at the peak of the crisis of contradiction about the nature of species.

Phillip R. Sloan commented as follows on the Buffon-Linnaeus controversy in a scholarly paper published in 1976:

> Linnaeus, whose long-standing policy was never to engage in public controversy with his opponents, made no public acknowledgement of Buffon's criticisms (other than naming the noxious-smelling *Buffonia* after him in 1748) [note date discrepancy], and proceeded with his systemization of nature as if Buffon had never existed, apparently deeming sufficient the reply given to all his opponents in 1748 that if his methods, "do not displease the Divine teacher of true method, I will welcome the fables of actors and barkings of dogs, with tranquility of soul."[2]

The answer to some of this confusion arrived with Charles Darwin a full century later. Darwin demonstrated that the differences between phyla, classes, orders, families, genera, and species were due to evolutionary descent through time. That explanation existed before Darwin's day, but it was not intellectually acceptable until after he had shown how it happened through natural selection.

Linnaeus became an evolutionist of sorts before his death at age 71. Apparently, the crisis of contradiction was something Linnaeus took to heart, for he abandoned the religious conviction on which he had once operated. The following two paragraphs were published in 1774, four years before Linnaeus's death.

> Let us suppose that the Divine Being in the beginning progressed from the simpler to the complex; from few to many; similarly that He in the beginning of the plant kingdom created as many plants as there were natural orders. These plant orders He himself, therefrom producing, mixed among themselves until from them originated those plants which exist today as genera.
>
> Nature then mixed up these plant genera among themselves through generations of double origin [hybrids] and multiplied them into existing species, as many as possible (whereby the flower structures were not changed) excluding from the number of species the most sterile hybrids, which are produced by the same mode of origin.[1]

Linneaus presents an outmoded creationist answer to Buffon's challenge that phyla, classes, orders, and families do not exist in nature. Evidence of common ancestry of plants and animals suggests that evolution began with a very few extremely simple forms of life. Buffon seems to have been more nearly right. The major classification divisions are just artifacts. We use them as models of the common characteristics between the long-lost ancestors of our constantly evolving species. Evolutionary change also explains the obvious variation between different members of every species. The whole problem of variation from an archetype is solved by establishing a more realistic definition.

George Gaylord Simpson and William S. Beck give the modern definition of species in their textbook, *Life*.[3] It is a condensation of a century of research set in motion by Darwin's work. Species are now considered to be

> populations of individuals of common descent, living together in similar environments in a particular region, with similar ecological relationships and tending to have a unified and continuing evolutionary role distinct from that of other species. In biparental species, the distinctiveness and continuation of the group are maintained by extensive interbreeding within it, and less or no interbreeding with members of other species. . . . The best and only *direct* evidence that an individual

belongs to a particular species is not the anatomy or physiology of the individual as such, but the observation in nature that the individual is living with the specific population and functioning as a member of that population. Such direct evidence is not available in practice for organisms that have been removed from the context of nature — that are specimens in collections rather than parts of living populations. Direct evidence is also lacking for fossils. In all of these cases pertinence to a systematic unit must be judged by indirect evidence. There are many kinds of indirect evidence, anatomy and physiology among them.

The one point to make here is that use of anatomy, for instance, as *evidence* that an organism belongs to a particular species does not mean that a species as a systematic unit is *definable* in anatomical terms. A species or other systematic unit is defined in terms of populations and their biological, evolutionary relationships. These relationships have anatomical consequences.[3]

We have examined the biological classification for animals and plants and found out two things. First, biological classification was never satisfactory until it conformed to the structure of nature. This is true of all classification systems throughout the sciences. Second, classification is a human construction. It reflects intellectual struggle. The one we have examined (Figure 3-2) is by no means complete, even though it has developed for nearly 250 years.

The Periodic Table of Chemical Elements

O Nature and O soul of Man!
How far beyond our utterance
are your linked analogies!
Not the smallest atom stirs
or lives on matter but has its
cunning duplicate in mind.[4]

Melville

Early Views of the Nature of Matter Herman Melville was quite right. Scientific models often seem to be little more than cunning duplicates of the mind. Nevertheless, some models are much better constructed than others. The periodic table of the elements is one of the best classification systems in all science. We can learn to appreciate it by contrasting it with older models. As each of these pioneering efforts failed, it narrowed the gap between an imagined structure of matter and what we now think of as its true form. Let's start with the Greeks of Aristotle's time (384–322 B.C.) and see what happened over the next twenty-two centuries.

At that time, all matter was believed to be composed of mixtures of five ingredients: earth, air, water, fire, and aether. (Aether was a mysterious substance breathed by the Olympian gods.) Gold, for example, was imagined to contain the fire of the sun, the malleability of water, and the massive quality of the earth. Alchemy, a folk pseudoscience that attempted to change base metals into gold, was practiced without success from ancient times until the development of modern chemistry in the late eighteenth century. Enthusiasm for it reached a peak near the turn of the eleventh century A.D. Alchemists mixed chemicals and metals together in anticipation of changing their properties to those of gold. They had ample reason to hope for success, because various smelting techniques were known that produced pure metals from very different-looking ores. It was not hard to imagine that the smelting fire added a sunlike quality while producing the change. These efforts were not entirely wasted. The alchemists were making experiments in a kind of primitive laboratory, and thus the science of chemistry was being born slowly, century by century.

By the middle of the seventeenth century, new ideas about the fundamental nature of matter began to be based on precise experimentation. In 1661, a British physicist, Robert Boyle (1627–1691), made an important contribution by defining a *chemical element* as a substance that could not be broken down into simpler substances. This idea opened the way to the study of *chemical compounds* that are made of two or more chemical elements.

Our modern definition of a chemical element differs from Boyle's in only two major respects. We now say that a chemical element cannot be broken down into simpler substances by normal chemical means. It is possible to break chemical elements down into various subatomic particles by forces that lie beyond the range of those in chemical reactions. Elements are also defined in terms of their subatomic particles. Each atom of each kind of chemical element must have the same number of *protons* in its nucleus, or center. (A *proton* is the unit of positive electrical charge found in the nucleus of the hydrogen atom, where it balances the negative electrical charge of a single orbiting *electron*.)

Our modern view of chemical compounds is also only slightly more specific than that of Robert Boyle. We now recognize that compounds are composed of two or more elements in fixed proportions. They exhibit invariable compositions no matter how they are made. Furthermore, compounds have unique properties that are readily distinguished from those of the chemical elements of which they are made.

By about the year 1700, science had reached a new level of sophistication built on the accuracy of instrumental measurements and technology. Ideas began to be expressed in quantitative terms. Alchemy gave way to the new science of chemistry, a science of discovering answers to questions of *how much* as well as just *what, which, how,* and *why.*

Elements, Atoms, and Atomic Weight Oxygen, hydrogen, and mercury were among the first elements to be identified. Each of them was obtained by breaking the mysterious chemical bonds that held it in compound form with other elements. Stronger bonds that defied attempts to break them concealed the existence of many other elements for a long time.

The concept of the *atom* as the smallest indivisible particle of an *element* was proposed in the middle of the fifth century B.C. by the Greek philosophers Leucippus and Democritus. The English chemist John Dalton (1766–1844) conceived the idea that different atoms and elements could be distinguished by the relative proportions in which they combined with one another in chemical compounds. At that time, it was not possible to weigh an atom on any mechanical scale because an atom is simply too small. However, Dalton saw the importance of knowing the *relative weight* of one atom of one element as compared to the relative weight of one atom of another element with which it could combine to form a molecule. (A *molecule* is the smallest unit of a compound that still retains its chemical identity.) Dalton's idea of obtaining atomic weights proved to be of tremendous importance. It enabled scientists to classify all the different elements and it opened the entire science of chemistry to mathematical analysis.

The key to the measurement of atomic weights was conceived in 1811 by an Italian physicist, Count Amadeo Avogadro (1776–1856). He postulated that equal volumes of all gases under the same conditions of temperature and pressure must contain the same number of atoms or molecules. Although Avogadro's hypothesis was not proved to be true until much later, it was recognized to be reasonable and was used as a law by chemists during the mid–nineteenth century. The idea became a very powerful tool. Here is the way it worked.

Laboratory experiments had already proved that two volumes of gaseous hydrogen held under standard pressures and temperatures were necessary to combine with only one volume of gaseous oxygen to form water. Using Avogadro's hypothesis, a nineteenth-century chemist could then conclude that every water molecule contains twice as many atoms of hydrogen as atoms of oxygen. Technology was sufficiently advanced by this time to permit weighing small amounts of hydrogen and oxygen in grams. After that, a simple mathematical ratio could be written and used to find the atomic weight of hydrogen, based on an arbitrarily assumed value of 16 for the atomic weight of oxygen.

$$\frac{\text{gr of hydrogen used}}{\text{gr of oxygen used}} = \frac{2 \times \text{atomic weight of hydrogen}}{\text{assumed atomic weight of oxygen}}$$

Imagine that a chemist has already found that 1.25 grams of hydrogen and 10 grams of oxygen form 11.25 grams of water. By substituting these numbers we can calculate the atomic weight of hydrogen as follows:

$$\frac{1.25 \text{ gr hydrogen}}{10 \text{ gr oxygen}} = \frac{2 \times \text{atomic weight of hydrogen}}{16}$$

Therefore,

$$2 \times \text{atomic weight of hydrogen} = 2$$

So,

$$\text{atomic weight of hydrogen} = 1$$

Oxygen combines with so many other elements that it can be used in the same way to find atomic weights of many different elements. The method was so simple and straightforward that it attracted a great deal of attention. Atomic weights for sixty-three elements had been measured by the mid-1860s. These numbers also began to attract attention. No two were identical. That was not particularly surprising but something else was.

Much had been learned about the chemical and physical properties of the different elements by this time. Many of the elements were found to share similar properties with one another and yet to have quite different atomic weights. Obviously, a classification scheme based on atomic weight alone made no sense at all. However, other classification clues were beginning to emerge from the information gathered by earlier workers.[5] As far back as 1829, a German chemist, Johann Wolfgang Dobereiner, had discovered several groups of three elements with similar chemical and physical properties but with strangely different atomic weights. Sulfur (atomic weight 32.06), selenium (atomic weight 78.96), and telurium (atomic weight 127.60) are one such group. At first, the important thing seemed to be that each number was almost double the value of the one before it. That was interesting, but was not the significant clue. Most important was that the numbers and physical and chemical properties suggested a classification symmetry.

Suppose all the different elements were lined up in vertical and horizontal rows on graph paper. The vertical rows could be based on chemical and physical properties. The horizontal rows could be based on rank of atomic number from left to right. It is a simple pattern as shown in Figure 3-4.

If things are really this simple there should be an element for every intersection of the horizontal and vertical lines. There was plenty of

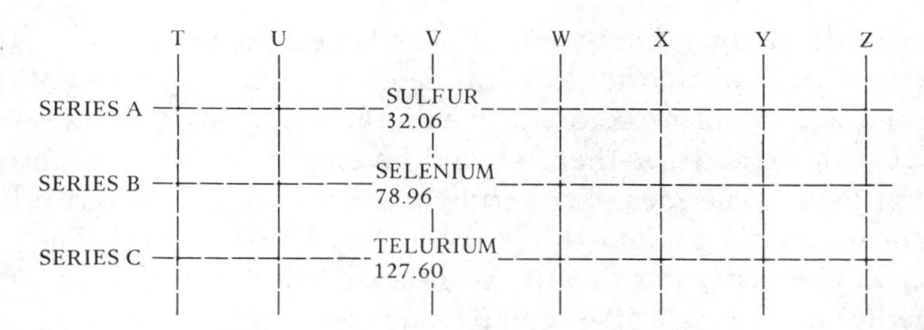

Figure 3-4 *This symmetry of classification of the elements was suggested by the work of Dobereiner in 1829.*[6] *He showed that some sets of elements with different atomic weights shared common physical and chemical properties. Sulfur, selenium, and telurium are an example. Imagine expanding on this idea while following the same symmetry by placing other sets of elements to the left or right depending on their atomic weights! (After Kleinberg, Argersinger, Jr., and Griswold)*

room left and right to accommodate all the known elements with their atomic weights set in advancing numerical value on each horizontal line.

Other bits of information made this a very appealing idea. The first, offered in 1857 by British chemist William Olding, reemphasized the idea that chemical elements occurred in sets like the sulfur, selenium, and telurium set discovered in 1829. Two examples of Olding's new sets are phosphorous, arsenic, and antimony; and chlorine, bromine, and iodine. Phosphorous (atomic weight 30.97), arsenic (atomic weight 74.92), and antimony (atomic weight 121.75) have atomic weights indicating that they must lie to the left of the sulfur set. Olding also recognized a set composed of chlorine (atomic weight 35.45), bromine (atomic weight 79.90), and iodine (atomic weight 126.90). Two of them, chlorine and bromine, exhibit atomic weights that justify placing them to the right of sulfur and selenium. Iodine has a slightly lower atomic weight than telurium and this presents a problem. Nevertheless, the idea has an enchanting appeal and would seem to be worth working out. Another bit of information was also developed by the research of a British chemist.

In 1864, John Newlands reported a "law of octaves" to the English Chemical Society. He had made a list of all the elements and ranked them in numerical order by increasing atomic weight.

Newlands thought that the physical and chemical properties of the first one were similar to those of the eighth, the properties of the second were similar to those of the ninth, and so on. If this idea was correct, it meant that there should be eight vertical rows on the graph paper. The idea is not truly accurate, but it did start other chemists thinking about the possibilities of such a chart. The science of chemistry was ripe for a synthesis, bringing all information together into a single all-encompassing idea.

A German chemist, Lothar Meyer (1830–1895), and a Russian chemist, Dmitri Ivanovich Mendeleev (1834–1907), although

Figure 3-5 *Lothar Meyer (1830–1895) was a German chemist who first recognized the possibilities of a full periodic classification of the elements in 1864. (Courtesy, MIT Historical Collections)*

Figure 3-6 *Dmitri Ivanovich Mendeleev (1834–1907) produced the most famous early periodic classification of the elements in 1869, revising it in 1871. (Courtesy of the Library of Congress)*

working independently, grasped the same great idea at almost the same time. Meyer was really first to publish. He introduced the idea of a periodic classification of the elements in a book in 1864. Mendeleev is more famous now because his 1869 paper entitled, "The Relation of the Properties to the Atomic Weights of the Elements" convinced scientists that this revolutionary idea was correct. It was a triumph. Both men had succeeded in developing the fundamental classification for all the chemical elements. Each continued to refine his ideas. Meyer published again in 1869 with a classification of fifty-six elements. Mendeleev followed two years later with another chart on which he classified sixty-six elements (shown in Figure 3-7).

Figure 3-8 is a modern version of the periodic table of the elements. This is obviously a complicated affair that cannot be understood fully without first learning basic chemistry. That much detail is beyond the scope of this book. Our approach must be a simple "Voilà!" There it is! Each set of elements that shares places in a vertical column is a set because they also share the same chemical and physical traits. There is nothing truly periodic about their variations except that each set is just a little different from the others in progressive steps of change.

Mendeleev's original table had blank spaces in it to accommodate elements that were as yet unknown. Nevertheless, he was able to predict what their properties should be. *Ekaboron* was the name he gave to one with a predicted atomic weight of about 45. This was discovered in 1879 by Milson and is now called *scandium*. Mendeleev named another undiscovered element *skaaluminum*. Its predicted atomic weight was about 70. This element was discovered by Lecoq de Boisbaudran in 1875 and is now called *gallium*. One more element, *ekasilicon,* was discovered within Mendeleev's lifetime. Winkler did it in 1886 and named it *germanium.* None of that seems particularly unusual until we examine exactly what Mendeleev was able to predict from what he knew about the structure of periodic behavior.

Prediction: The Crucial Test of a Good Classification System

The proof of the pudding is in the eating.

Cervantes

Figure 3-7 (facing page) MENDELEEV'S 1871 PERIODIC TABLE OF THE ELEMENTS[6]
Each vertical column is established on the way the individual elements shown as "R" combine with oxygen and hydrogen. For example, all the elements in column IV combine as RO_2, (that is, CO_2, carbon dioxide) and as RH_4 (that is, CH_4, methane). (After Kleinberg, Argersinger, Jr., and Griswold)

SERIES	I	II	III	IV	V	VI	VII	VIII
	—	—	—	RH_4	RH_3	RH_2	RH	—
	R_2O	RO	R_2O_3	RO_2	R_2O_5	RO_3	R_2O_7	RO_4
1	H = 1							
2	Li 7	Be 9.4	B 11	C 12	N 14	O 16	F 19	
3	Na 23	Mg 24	Al 27.3	Si 28	P 31	S 32	Cl 35.5	
4	K 39	Ca 40	— 44	Ti 48	V 51	Cr 52	Mn 55	Fe 56 Co 59
5	(Cu 63)	Zn 65	— 68	— 72	As 75	Se 78	Br 80	Ni 59 Cu 63
6	Rb 85	Sr 87	(?)Yt 88	Zr 90	Nb 94	Mo 96	— 100	Ru 104 Rh 104
7	(Ag 108)	Cd 112	In 113	Sn 118	Sb 122	Te 125	I 127	Pd 106 Ag 108
8	Cs 133	Ba 137	(?)Di 138	(?)Ce 140	—	—	—	— —
9	(—)	—	—	—	—	—	—	— —
10	—	—	(?)Er 178	(?)La 180	Ta 182	W 184	—	Os 195 Ir 197
11	(Au 199)	Hg 200	Tl 204	Pb 207	Bi 208	—	—	Pt 198 Au 199
12	—	—	—	Th 231	—	U 240	—	

IA	IIA	IIIB	IVB	VB	VIB	VIIB	VIIIB	
1 **H** 1.0079								
3 **Li** 6.941	4 **Be** 9.01218							
11 **Na** 22.98977	12 **Mg** 24.305							
19 **K** 39.0983	20 **Ca** 40.08	21 **Sc** 44.9559	22 **Ti** 47.90	23 **V** 50.9414	24 **Cr** 51.996	25 **Mn** 54.9380	26 **Fe** 55.847	27 **Co** 58.9332
37 **Rb** 85.4678	38 **Sr** 87.62	39 **Y** 88.9059	40 **Zr** 91.22	41 **Nb** 92.9064	42 **Mo** 95.94	43 **Tc** (97)[b]	44 **Ru** 101.07	45 **Rh** 102.9055
55 **Cs** 132.9054	56 **Ba** 137.33	57 **La** 138.9055	72 **Hf** 178.49	73 **Ta** 180.9479	74 **W** 183.85	75 **Re** 186.207	76 **Os** 190.2	77 **Ir** 192.22
87 **Fr** (223)[b]	88 **Ra** 226.0254[a]	89 **Ac** (227)[b]	104 **Rf** (261)[b]	105 **Ha** (262)[b]	106 (263)			

Lanthanide Series	58 **Ce** 140.12	59 **Pr** 140.9077	60 **Nd** 144.24	61 **Pm** (145)[b]	62 **Sm** 150.4	63 **Eu** 151.96
Actinide Series	90 **Th** 232.0381[a]	91 **Pa** 231.0359[a]	92 **U** 238.029	93 **Np** 237.0482[a]	94 **Pu** (244)[b]	95 **Am** (243)[b]

[a]Mass of most commonly available, long-lived isotope
[b]Mass number of most stable or best-known isotope(s)

Figure 3-8 *This is a modern periodic table of the elements. It is known as the short form because two of the horizontal groups have been shortened by extracting the Lanthanides (atomic weights 140.12 to 174.97) and the Actinides (atomic weights 232 to 260) from vertical column 3b.*

	IB	IIB	IIIA	IVA	VA	VIA	VIIA	VIIIA
								2 **He** 4.00260
			5 **B** 10.81	6 **C** 12.011	7 **N** 14.0067	8 **O** 15.9994	9 **F** 18.998403	10 **Ne** 20.179
			13 **Al** 26.98154	14 **Si** 28.0855	15 **P** 30.97376	16 **S** 32.06	17 **Cl** 35.453	18 **Ar** 39.948
28 **Ni** 58.70	29 **Cu** 63.546	30 **Zn** 65.38	31 **Ga** 69.72	32 **Ge** 72.59	33 **As** 74.9216	34 **Se** 78.96	35 **Br** 79.904	36 **Kr** 83.80
46 **Pd** 106.4	47 **Ag** 107.868	48 **Cd** 112.41	49 **In** 114.82	50 **Sn** 118.69	51 **Sb** 121.75	52 **Te** 127.60	53 **I** 126.9045	54 **Xe** 131.30
78 **Pt** 95.09	79 **Au** 196.9665	80 **Hg** 200.59	81 **Tl** 204.37	82 **Pb** 207.2	83 **Bi** 208.9804	84 **Po** (209)[b]	85 **At** (210)[b]	86 **Rn** (222)[b]

64 **Gd** 57.25	65 **Tb** 158.9254	66 **Dy** 162.50	67 **Ho** 164.9304	68 **Er** 167.26	69 **Tm** 168.9342	70 **Yb** 173.04	71 **Lu** 174.97
96 **Cm** 247)[b]	97 **Bk** (247)[b]	98 **Cf** (251)[b]	99 **Es** (254)[b]	100 **Fm** (257)[b]	101 **Md** (258)[b]	102 **No** (259)[b]	103 **Lr** (260)[b]

Mendeleev made these statements about ekasilicon (modern germanium) on the basis of the characteristics of the four elements that surround it above, below, right, and left on the periodic table.

> [Ekasilicon] . . . follows . . . the element silicon [in the table]. I named this element ekasilicon and its symbol Es. [Eka is a sanscrit prefix, meaning one, or one more.] The following are the properties this element should have on the basis of the known properties of silicon, tin, zinc, and arsenic. Its atomic weight is nearly 72 (actually 72.59), it forms a higher oxide EsO_4, a lower oxide EsO, compounds of the general form EsX_4 and chemically unstable lower compounds of EsX_2. Es gives volatile organometallic compounds; for instance $Es(CH_3)_4$, $Es(CH_3)_3Cl$ and $Es(C_2H_5)_4$, which boil at about 160°C . . . also a volatile liquid chloride $EsCl_4$, boiling at about 90°C and a density of 1.9 . . . the density of Es will be about 5.5 and EsO_2 will have a density of about 4.7.[7]

This quotation from Mendeleev is a magnificent example of prediction: the crucial test of a good classification system! We must be able to classify things as yet unknown as well as classify observations already made. There is only one state of nature. All things are within it. If we can predict, we know we have defined nature as it is. If we cannot predict, we know that our classification system is inadequate. Let us think for a moment at the student level.

Students tend to spend their thinking time getting ready for quizzes and attempting to predict what their professors will ask. This is equivalent to making a model of the discipline bounded by the limitations of the course. If the model is adequate, it will serve to predict what may be tested and will furnish a proper solution to any class related problem. We have already shown in Figure 2-9 how students should, and can, function as scientists. Prediction is a crucial test shared by all scientists.

Looking Back and Looking Ahead One of the most amazing things about this part of the history of chemistry is that the periodic table of the elements was developed without any knowledge of atomic structure. There was nothing theoretical about it. The whole process was empirical. Observations were made and compared. If something seemed to fit, it was retained. If not, it was discarded and something else tried. There is an important lesson here. Science is made as its practitioners search and find. That idea is implicit in our definition of science: *the progressive discovery of the nature of nature.*

Physicists have learned a great deal about the atom since the time of Mendeleev. Even so, electrons, protons, neutrons, neutrinos, positrons, quarks, and other subatomic particles, shells, orbital paths, and waves have yet to be fully understood. Once again, science is faced with a problem of proper classification that will parallel the structure of nature. We need a new kind of chart that will reveal the order of things and show where to look for the missing parts. There is an exciting human facet to this. Who knows where the next Lothar Meyer and Dmitri Mendeleev are at this moment? They could already be in the laboratory, or they may very well be evidencing their first childhood interest in science by collecting rocks, minerals, and butterflies. For all we know, they may be examining their baby rattles with unexpressed infant wisdom. They may be as yet unborn. They may be in your classroom. Whatever the story, we must wait for them.

Genius is known only by the fact accomplished.

Kenneth Frijoles

3-3 Steamboat Time: How Science Matures

Scientific progress is mirrored in technology. We can look at one and gain insight on how progress is made in the other. James Fitch and Robert Fulton were prominent American inventors who perfected the steamboat between 1787 and 1807. Their success depended on that of another inventor, James Watt, who had already perfected the steam engine. Watt, in turn, could not have made a proper steam engine until the machinery existed for fashioning the parts from well-forged iron. And so it goes; progress is a step-by-step affair. Each gain depends on a prior one. We will call this idea of progress "steamboat time."

The history of science shows the same stumbling development. Lothar Meyer and Dmitri Mendeleev could not have made their breakthrough until a sufficient number of atomic weights had been properly measured. Accurate laboratory work could not have been done until thermometers, pressure gauges, and chemical balances had been perfected and standardized. Perhaps the most fundamental part of their work depended on Avogadro's brilliant hypothesis. This idea made it possible to define the ratio of one element to another in a chemical compound. Earlier work on atoms, elements, and compounds was necessary before Avogadro could arrive at his insight. Creativity is a composite adventure. Each step aids the ones before it, but cannot stand without their support. The development of knowledge is a long, unbroken chain. Progress begets progress. But at what point is there enough progress to justify optimism that scientists are defining nature as it is, rather than making up imaginary views of it?

Alchemy led eventually to the science of chemistry, but alchemy was never chemistry. Alchemy was never a science. Its systems never paralleled the structure of nature. Here, then, is the key. *A science matures as its ideas can be shown to parallel and predict the structure of nature. Before that, a science is immature. Maturity begins with a classification system that synthesizes known facts and successfully predicts new ones.* Let us look at other illustrations that reflect the difference between scientific maturity and immaturity.

Plate Tectonics: An Idea and Classification System That Is Maturing the Science of Geology

We began discussing plate tectonics in Chapter 2, with Wegener's concept of continental drift. The geological details are complicated and need not be amplified further than to make just one additional point. Prior to 1961, geologists did not know how to link a wide range of individual observations into a single model of crustal behavior. The data were known, but no one knew how to classify them. At last the breakthrough came in a series of brilliant guesses. Alfred Wegener was vindicated! By 1968, geologists were able to look at almost any major structural feature on earth and classify it as part of a standard pattern.

The unifying theory came first and then produced its own classification system. The final phase of this scientific revolution occurred so rapidly that many able geologists were caught by surprise. This is the way one of them looked back on it years later.

> Those of us who received our geological education earlier in the century and who were involved in other facets of professional life were not fully aware of the enormity of the contradictions building up in the then current research just prior to 1960. Many of us still smiled at the suggestion of continental drift, confident that our professors had been accurate in claiming that there was no force capable of displacing such large parts of the crust. We still saw drift in terms of lumbering continental galleons impossibly plowing through sea floors of stationary Precambrian basalt. We still believed in the canon known as the permanence of continents and ocean basins. In this respect, we had ceased to think as scientists should think, for the act of believing is a part of a frozen judgement. Frozen judgements based on partial data are not compatible with the scientific ideal of growth of knowledge at the research frontier. We who thought this way were about to be re-educated by the true scientists among us, but we did not realize it at the time. That, scholars, is a confession.[17]

A Psychological Classification of Cognition and Perception: An Example of the Maturity of This Science

Perception of outside events by the human brain is an amazingly complex operation. We know very little about it. Psychologists are still trying to interpret behavior by analogy. The most impressive example of this is to be found in the subdiscipline of experimental psychology, where pigeons and white rats are used as laboratory animals. The idea is to discover why animals behave as they do in the hope of defining analogous human behavior. There are good arguments that justify this approach, but they do not change the fact that humans are neither pigeons nor rats. The complicated human brain must be capable of greater imagination and a wider range of reasons for making behavioral choices than are to be found in lower animals.

Donald M. Scott thinks that a healthy human brain behaves as if it is divided into three functioning areas.[18] A motor system controls muscles and coordination. An intellectual system controls cognition and perception. An emotional system controls emotions in part by regulating chemicals in the blood stream. These three systems operating together, feeding information back and forth between themselves, tell us what to do and ensure that we do it.

Psychologists wishing to study the intellectual system must examine the behavior of individuals under controlled conditions. This sounds scientific, but it cannot tell the whole truth. It is something like analyzing a computer's output to determine how it works without ever removing the plastic cover and looking inside. Nevertheless, for the present at least, this is the only valid way to study a living brain.

In some ways, psychologists face a task that is much more difficult than the one Mendeleev faced. He knew something about sixty-six of the one hundred or so chemical elements. Psychologists have no reason to feel confident that they are working with a representative sample of properly identified brain functions. This kind of uncertainty is typical of an immature science.

Let us examine one popular classification of one brain function to show the difficulties psychologists face. It was designed by William T. Powers. He concluded that a healthy human brain perceives and operates at nine different orders, or levels, of complexity, as shown in Table 3-1.[19] These nine orders of perceptual control represent a hierarchy of intellectual skills and feedback mechanisms that follow the input of signals to the brain. A young baby functions at the *first* order: it perceives input from its five senses and from internal irritants like hunger and fatigue. The brain is developed enough so that the infant knows when it is hungry, tired, hot, thirsty, feeling pain, in bright light, irritated by smells, and so on. At this order of perception, the baby is concerned only with the intensity of the signal.

A slightly older infant is capable of exercising the *second* order of perception. At this level, the baby must add inputs together and create more complicated sensations. For example, a baby can identify a rattle because it makes the proper noise when shaken. Brain development that has achieved the *third* order of perception is

Table 3-1
SUMMARY CHART OF THE NINE ORDERS OF COGNITION, PERCEPTION, AND CONTROL BY THE BRAIN

Order	*Description*
9th	Brain develops concept of the systems that give unity and direction to life; requires support of 8th order.
8th	Brain develops new principles on the basis of assembled data; requires support of 7th order.
7th	Brain performs intellectually on the basis of pre-existing and well-established known programs; requires support of 6th order.
6th	Brain functions with continuous conscious thought about changing spatial relationships; intellectual system directs motor system to interfere with 5th-order movements; requires support of 5th order.
5th	Brain directs motor system to function in sequences of movements; requires support of 4th order.
4th	Brain directs motor system to change one's own body position in space; requires support of 3rd order.
3rd	Brain capable of recognizing configuration, including position of one's own body in space; requires support of 2nd order.
2nd	Brain capable of "adding" inputs to create complicated sensations; requires support of 1st order.
1st	Brain capable of recognizing inputs from five senses and from interior of one's own body.

Sensory input occurs simultaneously at all orders but is not necessarily acted on by the brain

SOURCE: Adapted with permission, from William T. Powers' *Behavior: The Control of Perception* (New York: Aldine Publishing Co.). Copyright © 1973 by William T. Powers.[19]

capable of recognizing the spatial configuration of things, including the position of the child's own body in space.

All of us, young and old alike, do many things at these first three orders of perception. Suppose you have just placed your hand on a hot stove. The *intensity* of the heat is felt at the first order. At the second order, the *sensation* is interpreted as a potential burn. The threatening *spatial configuration* is recognized at the third order of perception. Higher levels of perception are necessary to do something about the burn. All these levels require conscious thought.

The *fourth* order of perception governs the decision to change position in space by dictating the *transition* of body from one place to another. This is the first level at which the brain's intellectual system interacts with the motor system directing it to get something done. Fourth-order perception is necessary before the hand will move away from the hot stove. This is an act of transition (from Latin, *transitio*, going over).

The *fifth* order of perception is much more complex. It involves *sequences* of physical movements. The brain controls the motions and is aware that they have taken place. The act of riding a bicycle is a good illustration. At the fifth order of perception, the brain would be able to direct sequences of transitions, or movement of the feet from one pedal position to another around and around. At this level of perception, the brain can also direct the motor system to keep the bicycle in balance. The act is one of maintaining sequences of motion. A good bicycle rider or automobile driver directs these sequences with little conscious thought. The brain seems to be programmed to function as if it is operating in "automatic pilot." Great athletes owe much of their seemingly effortless performances to well-developed fifth-order brain function.

The *sixth* order of perception entails continuous conscious thought about changing relationships. For example, the bicycle rider occasionally needs to decide whether to avoid a hole in the road or to stop at a cross street. These are conscious decisions and must be followed by motor responses. A bicyclist approaching a stop sign must make a judgment and act on it. Information on which to base a decision is displayed in the word STOP, the position of the sign in space, and the position of the cyclist in space. Failures in the sixth order of perception are very frequently observed at street intersections when two motorists attempt to occupy the same space at the same time.

The *seventh, eighth* and *ninth* orders of perception are totally intellectual. Muscle motion is not a part of them. The brain functions for the betterment of the intellectual component of the brain itself at these highest levels of cognition and perception. The purpose of the three highest levels of perception is to achieve not muscular control, which functions to operate the machinery of the body, but intellectual control.

The seventh order of perception is one of performing intellectually on the basis of pre-existing and well-established programs. In some ways, this is limited logical thought because it makes no demands on the brain other than to define which programs to apply in each situation. Standard solutions are known to fit. All we have to do is apply the right one. This thinking makes it possible to run a social organization like the army on the basis of standardized regulations. It also permits us to operate a nation by applying its laws.

The eighth order of perception is much more advanced for it is the level at which we develop new principles on the basis of assembled data. This is the first appearance of true intellectual creativity. Order eight extends across the human spectrum from the cunning development of strategies, to solving involved problems, to the writing of poetry.

The ninth order of perception controls a person's concept of *unity, direction,* and *order* in life. This is the way Powers explains the idea.

> Human beings seem to perceive unity in a collection of moral, factual, or abstract principles. On occasion they alter their choice of principles as a means toward achieving a more satisfactory sense of systematic unity, or to correct deviations of some perceived system concept from a preferred reference condition.[19]

Ideas about honesty, integrity, fortitude, and God are formulated at the ninth order. This does not mean that ideas like these are confined to this level. Many of them are organized as formal doctrines and used by followers at the dogma-strictured seventh order. Orders eight and nine are creative, whereas order seven is closer to the imitative world of monkey-see, monkey-do.

The higher orders of perception cannot exist alone. They must be accompanied by each lower order in turn. This is shown in Table 3-1. For example, the ninth order implies the existence of order eight to support it. Order eight must stand on the foundation of order seven. Seven, in turn, stands on six and so on down the line. The group is completely integrated. It represents a well-woven fabric of cognition and perception when seen from the top down. That is not surprising. However, a surprise is evident as soon as we begin to look at the classification system from the bottom up.

We have all seen children learning to operate at the lowest orders of perception. These actions are spectacularly displayed during the stage of development when infants first learn to stand and walk. We have seen them move their feet hesitantly, while they make sure of the position of each foot in space. By age 3, most children operate easily at the sixth order of perception. This requires a modest display of abstract intelligence. In fact, with the exception of input through experience, the sixth order is about the same level that is necessary to drive a vehicle, cut grass, or do most janitorial work. As we have already shown, the mechanics of athletic performance are developed at the fifth and sixth orders of perception. Athletes must practice, practice, practice to establish sequences of motions to the point that they no longer need to be performed consciously. Talent scouts say "a great player has the moves." Some of the highest-paid members of society earn their fees at the fifth and sixth orders of perception!

Most work of the world is done at the seventh order of perception, as both blue-collar and white-collar workers follow the standard codes of their employment. Anyone who is following instructions is operating at order seven when they make the decisions their work requires. The "brainy" professions like tax law are more difficult than serving as a postal clerk primarily because it is more difficult to master the governing codes. Obviously, seventh-order perceptions are carried out by people possessing a wide range of intelligence.

Eighth- and ninth-order perceptions require special creative intelligence. Fortunately, many people down through the milleniums have contributed to our lives in permanently creative ways. They have taught us to control fire, improve farm yield, speak our thoughts, to read, and to write. They designed machines and freed us from the limitations of muscle power. They were the philosophers who gave our lives purpose. The many contributions of their collective creative intelligence are obvious. Unfortunately, we are not at all sure how the human brain concentrates in order to be creative.

Sensory input occurs simultaneously at all orders, but it does not necessarily call for any action by the brain. Psychologists have not yet discovered how the brain amplifies certain signals and tunes out others. This quality may be the great "I choose" in each of us. Perhaps, there is an inborn habit that tells us, "It is all right to be arbitrary. Go ahead and select any perception you choose; amplify it and use it to define your own place in society." The consequences of such behavior may be the things that make some of us leaders and others followers.

Immaturity in a science is easy to see in the limitations of its classification systems. Nothing in the discussion of Power's ideas or in Table 3-1 considers the way a child learns to speak at age 2. Is speaking a combination of seventh- and eighth-order brain function or is it something else entirely? Psychology is still a science of elementary questions.

Classification and the Maturity of the Social Sciences

Economics, government, sociology, anthropology, archeology, and one branch of geography are the principal social sciences. Each of them is concerned with what happens as people interact with one another and with their environment. This makes them all behavioral sciences and therefore at least partially dependent on the development and use of psychology as a tool.

The dynamics of social groups are greatly affected by the behavior of their individual leaders. This in turn depends on individual brain function, perception, and choice. The mass suicide at Jonestown, Guyana, in 1978 is a frightening example; Hitler's control of the Third Reich is another. Steamboat time for the social sciences seems to be waiting for startling developments in psychology. In the meantime, social science research and classification efforts are slowly being developed as other sciences have been. A review of these disciplines will help us understand the problems they face.

Economics is concerned with the allocation of resources. Government, as a theoretical science, deals with political jurisdiction and the administration of public policy. Sociology, anthropology, and archeology examine slightly different facets of the origins, organization, institutions, and development of society. Geography is both a physical science and a social science. The part that is a social science studies the adjustments that people have made to the environments in which they live.

Controlled experimentation is difficult in the social sciences. As a result, most observations have been made on real-life situations, with real people making real choices according to what they believe to be their own self-interest. Real-life choices are rarely simple. Many factors influence them. Much of their complexity is due to the structure of society itself. Every individual belongs to several different organizations simultaneously. Each group may have its own competing influences, some obvious, others totally hidden. How would any sociologist be able to evaluate the effect of a childhood memory on some adult's willingness to behave rationally? Social units as fundamental as the chemical elements must be discovered before the raw facts are well classified and the social sciences can be called mature.

Although nothing has seemed to work so far, the search continues. Social scientists are very serious about their methodology and rigor. Their most elite practitioners do a lot of counting followed by sophisticated statistical analysis. Quantification is a powerful tool. Given the right data, it can distinguish important factors from unimportant ones. Quantification has been on the frontier of science for over two thousand years. That is why our next chapter is concerned with measurement and precision.

Annotated References

1. Jens Clausen, *Stages in Evolution of Plant Species* (Ithaca, N.Y.: Cornell University Press, 1951).
   Science is changing along with everything else. These quotations from Linnaeus demonstrate this change beautifully.
2. Phillip R. Sloan, "The Buffon-Linnaeus Controversy," *Isis*, 67 (1976), 356–375.
   The major part of this article deals with Buffon's recognition of the logical system of John Locke and how it might be applied to classification.
3. George Gaylord Simpson and William S. Beck, *Life — An Introduction to Biology*, 2nd ed. (New York: Harcourt, Brace, Jovanovich, 1965).
   Any scientist who wishes to obtain a better view of the problems of classification will find Chapter 18, pp. 488–493, very helpful.
4. Herman Melville, *Moby Dick: Or The White Whale* (New York: United States Book Company, 1892), p.545.
   At this point, with the ship becalmed, Captain Ahab pours out a soliloquy that seems to apply more directly to the preconception of modern science than to the religious passions with which he is struggling.
5. Heinz Cassebaum and George B. Kauffman, "The Periodic System of the Chemical Elements: The Search for Its Discoverer," *Isis*, 62 (1971), 314–327.
   This story is a complicated one. It shows the periodic system emerging slowly from the mists of many far-seeing, yet incomplete, efforts. The literature was scattered across Europe and inadequately distributed, so it is difficult to know who read what and when. It is also difficult to judge how much one author's work influenced another's. The periodic system was popularized primarily through the combined work of Mendeleev and Meyer. Contributions of men like Dumas, de Chancourtois, Newlands, and Olding will probably never be given full glory, because they stopped short of being sufficiently mind catching. There is a lesson in that. Great science must be mind catching.
6. Jacob Kleinberg, William J. Argersinger, Jr., and Ernest Griswold, *Inorganic Chemistry* (Boston: D. C. Heath, 1960).
   Mendeleev's periodic table is shown on p. 8.
7. Reed A. Howald and Walter A. Manch, *The Science of Chemistry: Periodic Properties and Chemical Behavior* (New York: Macmillan, 1971).

The quotation from Mendeleev occurs on pp. 5–6.

8. Robert S. Dietz, "Continents and Ocean Basin Evolution by Spreading of the Sea FLoor," *Nature* 190 (1961), 854–857.
   Dietz was the first person to publish this extraordinary insight.
9. Harry H. Hess, "History of Ocean Basins," in *Petrologic Studies: A Volume in Honor of A. F. Buddington,* eds. A. E. J. Engel, Harold L. James, and B. F. Leonard (New York: Geological Society of America, 1962).
   Once again, we see the coincidence of Meyer and Mendeleev grasping a great idea almost simultaneously, each working with separate data.
10. Allan Cox, Richard R. Doell, and G. Brent Dalrymple, "Geomagnetic Polarity Epochs and Pleistocene Geochronometry," *Nature*, 198 (1963), 1049–1051.
   Magnetic measurements proved to be the most important data developed since the time of Alfred Wegener. This information was anomalous enough to cause us to abandon the older belief in the permanence of continents and ocean basins.
11. F. J. Vine, and D. H. Matthews, "Magnetic Anomalies over Ocean Ridges," *Nature* 199 (1963), 947–949.
12. W. C. Pitman III, E. M. Herron, and J. R. Heirtzler, "Magnetic Anomalies in the Pacific and Sea Floor Spreading," *Journal of Geophysical Research*, 73 (1968), 2069–2085.
13. J. R. Heirtzler, G. O. Dickson, E. M. Herron, W. C. Pitman III, and X. LePichon, "Marine Magnetic Anomalies, Geomagnetic Field Reversals, and Motions of the Ocean Floor and Continents," *Journal of Geophysical Research*, 73 (1968), 2119–2136.
14. Patrick M. Hurley, et al. "Test of Continental Drift by Comparison of Radiometric Ages," *Science*, 157 (1967), 495–500.
15. Robert S. Dietz, and John C. Holden, "Reconstruction of Pangea: Breakup and Dispersion of Continents, Permian to Present," *Journal of Geophysical Research,* 75 (1970), 4939–4956.
16. Bryand Isacks, Jack Oliver, and Lynn R. Sykes, "Seismology and the New Global Tectonics," *Journal of Geophysical Research*, 73 (1968), 5855–5899.
   This is an extremely important paper, for it presents evidence to show how the movements that produce earthquakes may disclose the behavior of tectonic plates.
17. John T. Grope, personal communication.
   This is a pen name for a close associate of the author. Professor Grope has been a student and teacher of geology for many decades.

18. Donald M. Scott, personal communication covering material taken from an unpublished manuscript.
19. William T. Powers, *Behavior: The Control of Perception* (New York: Aldine, 1973).
Powers's ideas have been modified for presentation in a less technical context. The quotation came from pp. 171–172.

FOUR

MEASUREMENT: NEITHER MORE NOR LESS, BUT EXACTLY HOW MUCH

"When I use a word," Humpty Dumpty said,
"it means just what I choose it to mean —
neither more nor less."

Lewis Carroll

4-1 The Beauty of Quantitative Thinking Begins with Counting Things

The beauty of numbers is in their precision. They express exactly how much, neither more nor less. Numbers reveal relationships more clearly and more accurately than any other language. That is a strong statement, but it is true. Once numbers are correctly established, they eliminate all differences of opinion. Eight fingers are more than seven fingers. No amount of eloquence will change the established fact that eight is greater than seven. Counting things adds rigor to scientific reasoning by reducing the potential error in deciding what is and what is not least astonishing.

Suppose that we are interested in contrasting employment practices in economically developed countries with those in undeveloped countries. The United States of America and the People's Republic of China are good examples. A study of these two countries reveals a startling set of numbers. We may wonder about the source of the figures given in Table 4-1 and about the exact boundaries between categories, but we cannot ignore their impact.

Distribution of farm employment is by far the most surprising. Seventy-five percent of all the people gainfully employed in China work on farms; only 4 percent work on farms in the United States. This is a fundamental distinction, for it tells us something of the effort necessary to stay alive in these two countries. All animals, including humans are parasites that live directly or indirectly on energy that plants absorb from sunlight. Obviously, the Chinese have not been able to reap the plant harvest as efficiently as we do in the United States.

Farm employment in China is so high that only 15 percent of the workers are available to carry on trade, commerce, manufacturing, and other special services. The same group of occupations in the United States are carried on by 85 percent of the work force. The significance of these two figures is shown best by listing the occupations considered in the category "other special services."

These figures indicate that a well-developed economy places great emphasis on manufacturing, trade, commerce, and services. Raw materials on which these functions are based are obtained efficiently with a small manpower commitment. Underdeveloped countries exhaust their manpower resources in the effort to obtain

Table 4-1
PERCENTAGES OF GAINFULLY EMPLOYED WORK FORCE,
PEOPLE'S REPUBLIC OF CHINA VS. UNITED STATES

Occupation	*People's Republic of China*	*United States of America*
Agriculture	75%	4%
Manufacturing	8	23
Handicrafts	3	0
Mining	3	1
Construction	2	4
Trade and commerce	3	38
Transportation and communication	2	6
Other special services	4	24
Total	100%	100%
Total population	859,480,000	215,625,800
Gainfully employed	270,000,000	84,783,000
Percent of total population gainfully employed	31%	39%

These figures indicate that a well-developed economy depends on an efficient food source.

SOURCE: Adapted from Table 1.4, pages 26, 27, 230, Goode, John Paul, *Goode's World Atlas,* 15th ed., copyright © 1978, by Rand McNally College Publishing Company.[1]

enough food. The people who make life comfortable for the rest of us are the doctors, lawyers, preachers, teachers, artists, hair dressers, repairmen, cobblers, entertainers, civil servants, and military personnel. Imagine the price paid by the Chinese with only 4 percent of their gainfully employed population working in service jobs! The same category makes up 24 percent of the gainfully employed population of the United States.

Much of the wealth of the United States is drawn from manufacturing, trade, and commerce. These fields employ 61 percent of the workers in the United States and only 11 percent of the Chinese workers. That is quite a difference. Without manufacturing, trade, and commerce there can be little in the way of consumer goods available to the people. The United States was in this position in the eighteenth and early nineteenth centuries. At that time, our population was centered on the farms and forced to make many things for themselves. This is exactly what we saw in China as the decade of the 1970s came to a close. Science, aided by a new technology, especially the availability of abundant farm machinery, will put an end to the China we once knew.

The lesson here is not really one in economics. It rests with an understanding of numbers. Counting things gives reliable information and permits us to draw reliable conclusions. There is a formal beauty and uncompromising power in measurement. Eratosthenes (circa 276–195 B.C.) proved this when he single-handedly measured the circumference of the earth in about 215 B.C.

4-2 Eratosthenes Demonstrates the Formal Beauty and Uncompromising Power of Measurement

This is a story of imagination as well as an illustration of the beauty of measurement. Even now, 2,200 years after the event, it is hard to grasp the enormity of the feat. How was Eratosthenes even able to imagine making an accurate measure of a great ball 1.5×10^{23} times larger than himself? Perhaps part of our difficulty in grasping the reason for his insight rests with a cultural bias imposed by the common myth that Europeans thought the world was flat until sometime after A.D. 1492. That is nonsense. The spherical curvature of the earth had been known for at least two thousand years before Columbus sailed to America.

There are many proofs that the earth is round. Ships appear topsails first over the horizon and disappear hull down, topsails last. Travelers familiar with the star patterns of the middle latitudes see new patterns appear in the night sky as they move southward over the curvature of the earth toward the equator. The shadow of the earth cast in eclipse on the moon is curved. Observations like these gave the ancients an intuitive understanding of the facts, but science requires better proof and the certainty of measurement.

Eratosthenes found them in at least one source, the account of the explorer-geographer Pytheas (fourth century B.C.), a native of what is now Marseilles, France. Pytheas voyaged north along the coast of Europe exploring the British Isles and then sailed into the north Atlantic where he discovered an island he called Thule. Its location is still uncertain. All we know is that the description fits either Iceland or some part of Norway, where the summer days are so long that the sun "sleeps" only a few hours below the horizon. At different places in his travel, Pytheas measured the changes in the length of daylight and in the height of the sun at noon. These were accurate enough to be transcribed into measurements of latitude on a spherical earth. The method is shown in Figure 4-1. Notice that angles ϕ and θ vary with the position of latitudes L_1 and L_2. Pytheas made an additional contribution to map makers' views of the world by showing that the habitable regions extended far to the north before reaching the limit of the "frozen sea."[2]

All things have roots. It is easy to see how the work of Pytheas influenced Eratosthenes because Thule, Britania, and Ierne (Ireland) were placed on his own map of the world. Other sources of ideas and inspiration are more difficult to assess. Eratosthenes had an excellent education. Four of his teachers in Athens were well-known followers of Aristotle. A fifth, Bion the Cynic, offers tempting ground for speculation about possible relationships between creativity and lifestyle. Eratosthenes left Athens when he was about thirty years old to take a post in Alexandria as tutor in the court of Ptolemy III, pharaoh of Egypt. He became head of the great library and museum about ten years later, holding the position for about

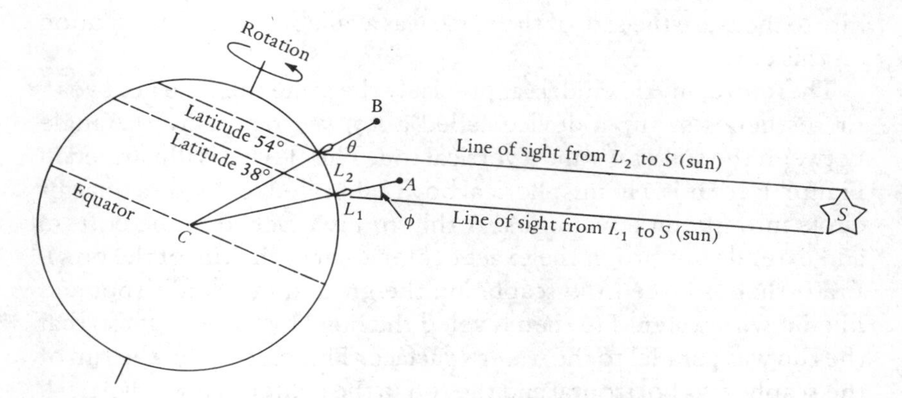

Figure 4-1 *The sun angles Φ and Θ measured between a vertically held weight on a string and a line of sight to the sun vary with the latitude of the observer. Pytheas made observations like these as far north as Thule (Iceland?) during the fourth century* B.C. *He also noted the changes in the hours of daylight that occur with what we now recognize to be positions of latitude on a rotating earth.*

forty years until his death. This was no ordinary opportunity for a scholar in those days. Alexandria had the finest library and research institution in the world. Intellectual stimulation must have come from scores of visiting scholars drawn to Alexandria by the chance to share these resources. We know that Eratosthenes was involved in many projects. He was a poet, a grammarian engaged in standardizing language, a literary critic, an historian involved in establishing the dates of past events, a mathematician interested in prime numbers, and a geographer. His peers are supposed to have nicknamed him "Beta," signifying the dull B student, hard working but poorly endowed. Such opinions are of no consequence against the single sign of true genius, the accomplished fact. Here is his grand conception.

Eratosthenes knew that there was a deep-water well near Syene at the first cataract of the Nile some five thousand stades (a distance unit) south and slightly east of Alexandria. (See Figure 4-2a.) This water well was almost exactly on what is now called the Tropic of Cancer. At this point on the longest day of the year, the noon sun shone vertically down into the well and was reflected in the water at the bottom. Eratosthenes realized that at this moment a straight line would connect the sun, the well, and the center of the earth as shown by line *SWC* in Figure 4-2b. He realized that the opportunity to measure the size of the earth was available every year at noon on this day.

Therefore, in Alexandria at precisely the same time the next year, Eratosthenes set up a device called a *scaphe* to measure the angle between the sun's rays and a vertical rod. This device is illustrated in Figure 4-2c. It is a hemispherical bowl calibrated to divide a quarter circle into fifteen equal units. A thin rod is attached to the bottom and extends upward in the exact center as far as the rim of the bowl. Eratosthenes placed the scaphe on the ground in a sunlit spot and filled it with water. He then leveled the bowl by adjusting it so that the rim was parallel to the water's surface. This meant that the rim of the scaphe was horizontal and the rod in the center of it was vertical. He now had a line along the length of the rod that pointed downward directly toward the center of the earth. His next step was to drain away the water so that the tip of the rod, *R*, could cast a shadow on the inside of the bowl. From this he could read the value

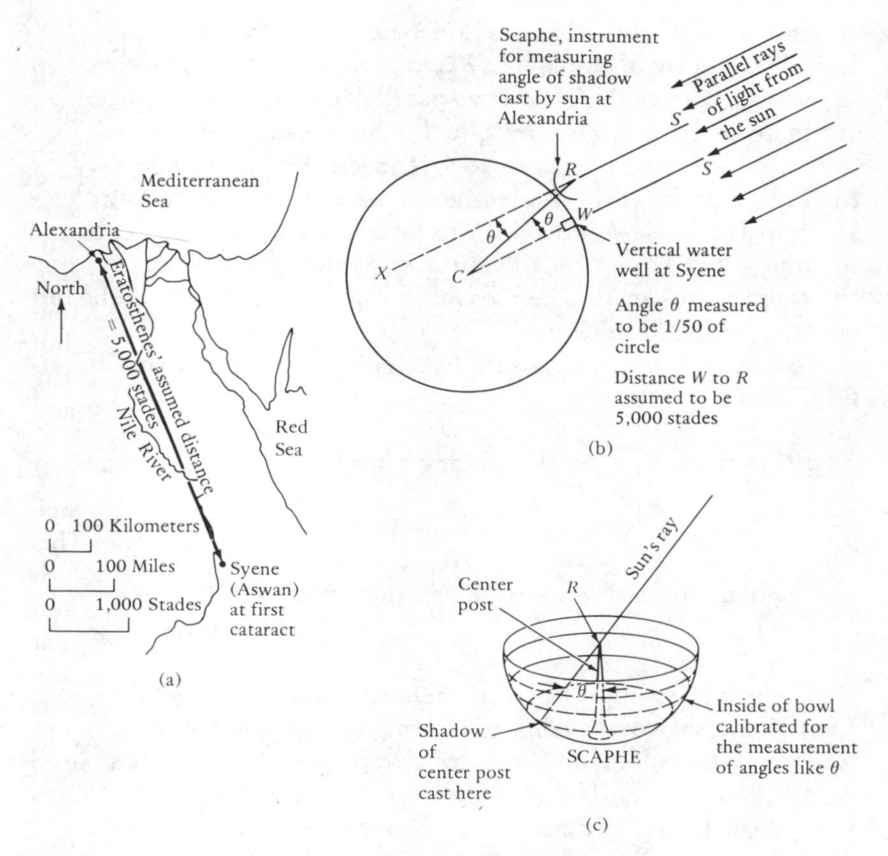

Figure 4-2 *Eratosthenes measured the polar circumference of the earth in about 215 B.C., using a base line between Alexandria and Syene, Egypt, as an arc length. Parallel rays of sunlight gave him a measurement of the angle at the center of the earth that subtended his baseline arc. Simple multiplication furnished the value of the circumference.*

of angle θ and thus measure the angular height of the sun, as shown in Figures 4-2b and 4-2c.

We have reached a very important point in the illustration. Do not trust memory. Look back at Figure 4-2b. Eratosthenes appreciated that sunlight is cast toward the earth in parallel rays. Therefore, line *SRX* from the sun to the tip of the rod and into the bowl could be extended into the earth as shown. The elegance of his idea should become clear. He has constructed two parallel lines *SRX* and *SWC* cut by diagonal line *RC*. There is a well-known theorem in plane geometry that when two parallel lines are cut by a diagonal, alternate interior angles are equal. This means that angle *XRC*, measured in the scaphe as angle θ, is exactly equal to angle *RCW* at the center of the earth. Eratosthenes had managed to measure the value of the angle θ, now seen to be at the center of the earth! It separates two points on the surface, Syene and Alexandria, five thousand stades apart. The measured value of θ was 1/50 of the full circle, or in modern terms 7°12′.

He was now left with a simple calculation of the polar circumference.

$$5{,}000 \text{ stades} = 1/50 \times \text{polar circumference of the earth}$$

So,

$$\begin{aligned}\text{polar circumference of the earth} &= 50 \times 5{,}000 \text{ stades}\\ &= 250{,}000 \text{ stades}\end{aligned}$$

All we need to know now is the length of a stade and we can check Eratosthenes's work against modern values for the polar circumference. Unfortunately, there are several possibilities, each based on different standards under different political systems. A Roman stade is the length of a stadium and equal to 185 meters (607 feet). An Athenian stade is equal to 177 meters (582.7 feet). An Egyptian stade derived as 1/32 of another measure, the schoenus, is equal to 157.5 meters (516.7 feet). Any of them could be the value that Eratosthenes measured. Let us look at each of them in turn.

250,000 Roman stades	=	46,400 km (28,800 miles)
250,000 Athenian stades	=	44,400 km (27,600 miles)
250,000 Egyptian stades	=	39,400 km (24,500 miles)

The modern value for the polar circumference of the earth is very close to 40,000 kilometers (24,900 miles). If Erathosthenes's measurement was made in Egyptian stades it was 1.6 percent too small. In Roman stades it was 15.6 percent too large. In Athenian stades the measurement was 10.8 percent too large. All these figures are remarkable considering the instrument he used and the limited information at his disposal.

The method is almost perfect. A small error is introduced because Alexandria is not exactly due north of Syene. Another source of potential error lay in the position of Syene, which is not quite on the Tropic of Cancer, so the sun was not truly overhead when he made the measurement. This would affect the value of Θ. A third error rests with the measurement of the length of the baseline. It should be 869 kilometers (540 miles). Using Eratosthenes's value for θ and the proper length of the baseline, we may calculate a polar circumference to be 43,500 kilometers (27,000 miles), with an error of 12 percent.

All these figures are interesting but not critically important. The important fact is that Eratosthenes devised a way to measure the diameter of the earth and to dispel for all time the idea that the "known world" was the whole world. That lesson was not fully appreciated before Columbus. *Here at last, we have the true substance of science: the chance to see it all and to see it free from cultural biases.*

> When his small son asked Einstein why he was famous, the father replied, "You see, son, when a blind bug crawls along the surface of a sphere, it does not notice its path is curved. I was fortunate enough to notice this."[3]

4-3 Standardized Units of Measurement

Standardized units of measurement were established in antiquity for the practical purposes of trade and commerce. Units of weight, length, area, and volume were first formalized by the authority of local custom and then backed by the laws of governments. This system is satisfactory unless ambiguities develop. Eratosthenes used the word *stade* without citing the proper government standard: Roman, Athenian, or Egyptian. It is reasonable to imagine that he used the Egyptian stade because he did the work in Egypt and was on the pharaoh's payroll. Nevertheless, we do not know.

An ambiguity like this is unsatisfactory by modern standards. Science is now international in scope and requires internationally accepted units of measurement.

The British Imperial System and the United States Customary System are both inadequate for scientific use. They grew haphazardly from many unrelated sources and contain hopeless ambiguities. For example, two kinds of quarts exist in the U.S. system. A liquid measure quart contains 57.75 cubic inches and a dry measure quart contains 67.2 cubic inches. A U.S. barrel may contain anywhere from 31 to 42 gallons, depending upon the legal regulations governing the kinds of materials to be measured. Fortunately, a genius appeared at the very birth of modern science and offered a solution to the problem of standardization.

The Metric System

In 1670, Gabriel Mouton, Vicar of Lyons, proposed the Metric System of Weights and Measures to the French Academy. People react slowly to suggested changes in the way they should do things. More than a century was to pass and an entire political structure torn down in the French Revolution before the metric system was adopted by the National Assembly in 1795. Seventy-one years later, in 1866, the Congress of the United States granted the metric system legislative sanction. Scientists throughout the world now use it almost exclusively. American and British engineers use both systems, but tend to think in pounds, feet, miles, and gallons. Kilograms, kilometers, and liters seem to be beyond the comprehension of the English speaking public. Soft drinks in liter bottles are often thought of as "funny quarts". Nevertheless, cultural biases must eventually give way to practicality. There is a persistent beauty in the metric system that will impel us to join with the rest of the world and accept it.

The most obvious beauty of the system is the way it is structured mathematically on powers of ten. For example, the *meter* is the principal unit of length. It was originally conceived to be one ten-millionth (1×10^7) of one-quarter of the circumference of the earth as measured along the longitude meridian that passes through

Paris, France. That proved to be much too indefinite a figure for modern usage, so the meter was redefined in 1960 as the length equal to 1,650,763.73 wave lengths in vacuum of the orange-red radiation of krypton 86. A *centimeter* is 0.01 meters. A *millimeter* is 0.001 meters. A *kilometer* is 1,000 meters. This regularity continues throughout the system. Units of volume are measured in cubic centimeters and liters. A *liter* contains 1,000 cubic centimeters. Units of *mass* reflecting the amount of matter in a body are measured in *grams* and *kilograms*. Once more, the larger unit (the kilogram) is 1,000 times greater than the smaller one (the gram). Units of temperature are measured in *degrees Celsius*. They are based on one hundred divisions between 0°C at the freezing point of water and 100°C at the boiling point of water. The scale is standardized at sea level under one atmosphere of confining pressure. Units of force (weight) are also organized in powers of ten. The smallest unit is the *dyne* (measured in g cm/s^2). The principal larger unit of force is the *newton* (measured in Kg m/s^2).

Gabriel Mouton was a genius, for his metric system reached to the very heart of science. It allowed for the progressive discovery of the nature of nature. The discovery of the true nature of weight is a particularly good illustration, for it shows how the system can be expanded and yet remain internally consistent. We are already familiar with a few measurements like those of time, length, and volume that are obtained directly by dividing the unknown into standardized parts and counting the units. This is the way we count the number of ducks in a flock. More complicated measurements must be derived. Weight is such a quantity. For thousands of years, our predecessors had the mistaken idea they were weighing things using a beam balance, when they were actually determining *mass,* not *weight.*

The Contrast Between Direct and Derived Units of Measurement

When an ancient goldsmith measured what was then called the "weight" of some quantity of gold on a beam balance, he or she must have appreciated that the numerical value was a property of a specific amount of the metal. The method was simple. An unknown quantity of gold was put in one pan and balanced by a certain number of standard weights in the other pan.

What was not known in ancient times is that true *weight is a force* due in part to the *gravitational acceleration* of the earth and in part to the *mass* of the object. These concepts must be defined. *Force* (symbol, F; unit, a dyne, g cm/s^2) is a push or a pull capable of moving and accelerating a free body. *Acceleration* (symbol, a; units, cm/s^2) is the technical name for the rate of change in *velocity* (symbol, v; units, cm/s) as a free body is moved uniformly and ever more rapidly under the continuous application of a force. *Gravitational acceleration* (symbol, g; units, cm/s^2) is the acceleration given to a freely falling body impelled by the force of gravity. The value of g that is usually used is 980 cm/s^2. *Mass* (symbol, M; unit, g) is a measure of the amount of matter in a body. A mass of 3 grams has three times as much matter in it as there is in 1 cubic centimeter of water.

Sir Isaac Newton (1642–1727) was the first person to grasp the significance of the true relationship between *weight, mass*, and the *acceleration of gravity*. Newton recognized that a beam balance measures *mass* rather than *weight*. The balancing cancels out the effect of gravity pulling downward in the same way on both sides of the balance. Mass can be measured directly like length, volume, and time. Weight must be derived through this equation.

$$\text{weight (a force, } F) = \text{mass } (M) \times \text{acceleration of gravity } (g)$$

Or,

$$F = M \times g \text{ (units of force in dynes are, g cm/s}^2)$$

Most scientific measurements are expressed in derived units as products or ratios of other things. Many of them are very familiar to us in the form parts *per* million, force *per* unit area, miles *per* hour, or vibrations *per* second. Measurements like these carry with them a special message telling us exactly why they were developed. The word *per* notifies us that the field measurements (parts, force, miles, and vibrations) represent large undivided populations of things that

Figure 4-3 *Sir Isaac Newton (1642–1727) was the first person to recognize that weight is a force equal to the product of a property of matter called* mass *and the acceleration of gravity. (Courtesy of the Library of Congress)*

need to be compared on a more limited basis to be understood. A sound wave vibrating at 440 cycles *per* second defines the tone we call "A above middle C." Comparison increases the usefulness of precision provided that the reasoning behind it is understood.

Beginning science students often do not understand derived units of measurement. We are back to the ideas expressed in Figure 2-9, comparing the fabric of science as perceived by professional practitioners and by students. It is frustrating to work with measurements that do not make sense. That is unnecessary if the origins of all derived measurements are thoroughly investigated before they are used.

The Rationale for Using Derived Units of Measurement in Psychology and the Social Sciences

Psychologists and social scientists are concerned with people's behavior both as individuals and as populations of individuals. Behavior is acting on choices after gathering information and making decisions based on an estimate of the situation. Research in this area is defined in one set of questions: *who makes what and which choices; when, where, how, and why?* Each science has its own ways of approaching research questions, but they all share a common starting point — measuring and describing human behavior.

Measurement should be the most accurate form of description because it is quantitative rather than literary. Unfortunately for the development of the social sciences, it is very difficult to measure human behavior and to know exactly how each action is related to the information on which decisions were based. More progress in psychological research will be necessary before social scientists are able to overcome this limitation. This does not mean that other sciences must stand still until psychology matures. There is a way out and social scientists are taking it.

Economics professors Armen Alchian and William R. Allen have explained their technique before building models of economic systems.[4] They approach their science as a game founded on six postulates that function like the axioms in plane geometry. A *postulate* is something that is taken for granted and used as if it is true. No defense or claim of accuracy is made for a postulate. It is simply used to see where a hypothesis might lead. Everyone is encouraged if the results seem promising and without major contradictions. The test of this game is a pragmatic one. The ideas must work or be discarded and replaced by some other ideas that will work. This is equivalent to saying, "Some day psychologists may discover that these six things are true."

1. For each person some goods are scarce.
2. Each person desires many goods and goals.

3. Each person is willing to forsake some of an economic good to get more of other economic goods.
4. The more one has of any good, the lower its personal marginal valuation. [*Personal marginal value* is the increment of value placed by any person on the possession of one additional unit of something.]
5. Not all people have identical tastes and preferences.
6. People are innovative and rational.[4]

Two more "unofficial" postulates might be added to this list and used to broaden the base to cover all the social sciences.

7. People always take what is for them the most rewarding course of action.
8. People tend to follow the dictates of the cultural patterns around them.

Explanations of postulates 7 and 8 are probably necessary to avoid confusion with the more authoritative ideas of professors Alchian and Allen. All new individual actions, whether of thought, speech, or movement, require some change from an earlier state to a new one. The decision to do something is established as a "go" or "no-go" situation. Enlightened self-interest demands that an individual see some impelling benefit before selecting one action over another. This is true even if the choice results in seemingly unpleasant consequences. At a trivial level, one might choose to attend a funeral rather than play golf. Such duties often preserve an image as a first-class human being. Choosing the pleasant alternative could be a public admission of second-class status.

Postulate 8 is justified because it is almost unnatural for any individual to be constantly at war with the cultural environment. People surrender to the dictates of their learned and shared ways of doing things as a matter of expediency.

An Example of a Derived Unit of Measurement in Economics
Gross national product (GNP) is defined as the market value of all final goods and services produced in a nation's economy during a given time, usually a calendar year. In the United States, the GNP is calibrated in dollars per year (for example, $/1977). The figure is usually given in billions of dollars.

Derivation of the GNP is an interesting exercise. Market value varies from year to year. The cost of an item changes with fluctuations in supply, demand, and inflation. For this reason, the actual numerical values of the GNP have little meaning until compared with one another against some standard, perhaps the state of the economy in 1969. Quantification is used to permit precise description, yet practitioners in this social science do not agree on an absolute standard.

The GNP is compiled from four categories of goods and services: consumption goods, gross domestic private investment, net exports, and government purchases. *Consumption goods* are all goods and services purchased by individuals for their own use and pleasure. This category includes necessities like food, medical services, utilities, gas and oil for automobiles as well as recreation, such as at theaters and clubs. *Gross domestic private investment* covers expenditures for inventories, new machinery, labor, and so on. *Net exports* is a category that totals the difference between the value of all exported goods and services and the value of all imported goods and services sold here. We sometimes speak of it in terms of the balance of payments into and out of the country. *Government purchases* include the value of all things purchased by the federal, state, and local governments. Costs of maintaining the armed services, public education, buildings, and roads are included under this heading.

The GNP is compiled on the final values at the time of consumption rather than at each step during manufacturing. Nothing is counted twice. For example, the value of one new automobile includes the value of each part and the labor necessary to produce it. The measurement would be in error if separate tallies were made for the cost of the iron ore, the raw rubber, or glass sand. GNP is an appropriate name to describe this measurement.

The nation functions as a complex population of individuals interacting within the rules laid down by the eight postulates. There is something majestic about summing up all this activity and defining its value as a single number. The idea is enough to make an economist smile with pride and a humanist twinge.

An Example of a Derived Unit of Measurement in Psychology

Most tools of modern science are used by psychologists as they study brain functions and related behavior in both animals and humans. One example will show how a "go, no-go" behavior in rats may give insights on human behavior. This example symbolizes all scientific research, in which way leads to way and progress leads to new problems as yet undreamed.

A typical experiment is performed on a few pigeons or white rats separated from one another and treated individually in isolation chambers. Each animal is presented with a standard set of stimuli to which it responds for an hour or so every day. Measurements are usually based on automatically recorded responses per units of time. At the end of one or two months, all the animals have been trained to respond in identical fashion. This standardized behavior represents a baseline from which future deviations may be measured.

The experimenter then introduces a new set of stimuli and measures the changes that result from this intrusion on well-developed habits. Analysis of the data is difficult unless the experimental design eliminates all reasonable ambiguity as to why the measured changes have occurred. The threat of a *post hoc ergo propter hoc* fallacy *(after this, therefore because of this)* is an inevitable difficulty in experiments of this type. The best experimental designs produce identical behavior patterns in all the test animals. These identical responses eliminate the chance that each animal is performing on the basis of its own whim rather than as a result of the uniformly applied experimental stresses. Experiments of this type are taken to be definitive when they can be successfully repeated with different sets of animals and always yield the same results.

In one experiment, the psychologists wanted to determine whether or not large doses of alcohol change the way rats perceive time.[5] This is basic research, but it is easy to see far-reaching implications. Many animal studies correlate with humans because animal anatomy is related to human anatomy through organic evolution. Four "drunk" 120-day-old male albino Holtzman rats, measuring time and pushing levers, are not teenage humans driving automobiles while under the influence of alcohol, but similarities do exist.

The rats were first thinned down to 80 percent of their normal free-feeding weight. This was done to ensure that they would be hungry enough to exchange physical work for food. (See Alchian and Allen's economic postulates 1, 2, 3, and 6, pages 114–115, and the intentional design to suppress the effect of postulate 4.) The rats were then placed in isolation boxes, where they learned to press a lever and receive one standard food pellet. As soon as a pellet fell into the feeding bin, a timing device was set to penalize any rat that attempted to obtain another pellet before the end of a twenty-second interval. If a rat pushed the lever again during this time, the clock was reset and started to measure a new full twenty-second period. The rats soon learned to wait patiently for about twenty-one or twenty-two seconds before attempting to eat again. All the rats became adept at measuring the standard time interval accurately by the end of a fifty-day training period. A baseline was considered to be established when the ratio of payoffs to lever-pressing efforts varied no more than 10 percent for four consecutive days. The psychologists were then ready to introduce rats to large doses of alcohol injected directly into the abdominal cavity.

Each rat was tested with varied concentrations of distilled water and alcohol. These ranged from pure distilled water to concentrations between 0.43 and 1.7 grams per kilogram of animal weight. The scale was equivalent to a range of 2, 4, and 6 traditional 1.5-ounce drinks for a 150-pound (68-kilogram) human. Care was taken to rest the rats between experiments. They were allowed four days in which to recover before receiving another injection. During the rest period, each rat continued to respond normally to the routine training schedule.

No spectacular results were obtained until the alcohol dosages reached the highest level. Then the rats' abilities to measure time were severely altered. They responded too quickly by a factor of about 22 percent of their standardized behaviors.

The experiment suggests, above all, that large doses of alcohol may interfere with any habitual behavior pattern in which time is important. Imagine what this could mean for automobile drivers who feel comfortable driving at high speeds. A 22-percent shortening of time perception could mean that such a driver feels as if the car is moving much slower than it is. This in turn might lead to what could appear to be vainglorious bravado or an inability to judge braking time. Animal experiments have much to tell us that we need to know.

A Critique of Derived Units of Measurement in Sociology and Political Science In the late 1860s, a young New Hampshire schoolboy named Plupy Shute was given a simple lesson in sociology by his father. Papa Shute divided the population of the United States into four social classes:

1. *The Bobs* Poorly educated unskilled laborers and their families
2. *The Tippy Bobs* Moderately educated shopkeepers, skilled tradesmen and their families
3. *The High Tippy Bobs* Professional educated people of all ranks: doctors, lawyers, preachers, teachers, top politicians and their families
4. *The High Tippy Bobs Royal* Financially endowed people who do not do anything and their families that help them do it

More than a century of serious research has passed since Papa Shute provided this humorous summary of society. Nevertheless, sociologists and political scientists are still trying to discover the fundamental categories into which society is divided. Research is quantitative now. A wide range of units have been developed, most of which involve counting things. For example, demographic tables defining the state of world population tell us all sorts of things. We have counted the households in Samoa with more than ten people in them (3,228).[6] We know that there are 210 elected black officials in Mississippi and only 4 in Nebraska.[7] The numbers are known, but they are static whereas society is dynamic.

Such information describes but does not explain. Consider that all demographic data are probably reasonably accurate. Each bit of information defines one aspect of the fabric of society in cold and passionless terms. The *Demographic Yearbook for 1976* can tell us how many women in Mexico are divorced.[6] The number describes some part of society, but it is distilled from the acts of living. All social scientists face a difficult paradox. We understand society at the small-group level, yet find it confusing at larger levels. Each of us understands the behavior of some part of our family and a very small group of people about us. We know a great deal about why these individuals behave as they do most of the time. This is true because we interact with them all on a one-to-one level and can watch them interact with other people at the same one-to-one level. Beyond that range, when the bonding forces weaken, we understand much less.

Sociology and political science might profit from the development of a new paradigm. It is possible to conceive one in which the fabric of society is structured as it is in real life on the strength of bonds between individuals. Elements of it function like the bonded parts of a great water molecule that constantly passes through itself in slithering, weaving chains and threads, mad snakes that temporarily flow this way or that way as the pressures on them impose stress beyond their strengths to resist. All of nature is in *dynamic* equilibrium, a state of change in which the kinds of change and rates of change are proportional to the forces involved. The bonding strength of the parts define the segments that keep their shapes and move as units. Bonding strength seems to be the factor that must be measured before the larger elements of our social order make sense.

4-4 Learning to View Things in Proper Scale

Nature is difficult to comprehend. There is so much of it. To gain perspective, we need to find ways to view nature in large segments without losing too much detail. We can construct quantitatively accurate models that are small enough to be seen, yet accurate enough to represent the real world. Model building leads to many surprising insights. Here are a few sample models to encourage you to make others.

Personal Scales for Measuring Time and Distance

Human history is complex. Most of it seems unreal to us and fades away in the remoteness of past time. Dates are hard to remember. Those we do know, like 1776, 1492, 1066, and 55 B.C., are so far apart and spaced so irregularly that they are inadequate to calibrate the steady flow of world history. There is one startling exception to this generalization. We all seem to appreciate the reality and sequence of things that have happened in our own lifetimes. This suggests that each of us has a personal scale by which we measure historical time.

In this scaling system a twenty-year-old college student has lived a little more than 1 percent of the time since Christ was born. The American Civil War took place only six student life spans before the present. Wonder of wonders, Pickett's charge took place only two student life spans before the miracle at Kitty Hawk. What a change occurred in those forty years! It was as spectacular as the changes that followed World War II.

A sixty-year-old person has a longer span of time to use as a personal scale. World War I is almost within memory. To such a person, the outbreak of the American Civil War is only two units back and one of those is filled with living memories. The rediscovery of America by Columbus took place eight life spans before the present. An even more startling realization is that a sixty-year-old person has lived a full 3 percent of the time since the birth of Christ. No wonder older people are aware of the value of their experiences. Many things have happened to them and around them. They have seen history unfold.

Geological time, or earth time, began 4.5 billion years ago. This can be scaled off in the same way, but the very vastness of 4.5 billion years makes life-span units meaningless. After all, sixty years is a very small part of 45 million centuries. The key to appreciating geological time rests with the ability to look at rocks as records of events and with the imagination to restore ancient mountain ranges, rivers, floods, sea floors, waves, currents, and winds. The process is not magical. We all do similar things on a daily basis. Think of the number of times we have admired some charming elderly woman and thought, "Wow, what a beauty she must have been when she was nineteen!" Try the same imaginative approach on the picture of the Grand Canyon, Figure 4-4.

The Grand Canyon is about 8 million years old. Before that time, the Colorado River flowed across the plain marked by the straight line of the far horizon. The inner gorge at the bottom of the canyon has been cut in crystalline rocks *(X)* that are about 1.2 billion years old. These rocks are the last remaining roots of a chain of mountains that had been eroded to a flat plain 600 million years ago. The

Figure 4-4 *The Grand Canyon is an excellent place to learn to look at rocks as records of separate events occurring over time. With imagination, it is possible to reconstruct the sequence of things that happened here throughout more than 1.2 billion years. The lowest rocks* (X) *indicate an eroded range of Alpine mountains. Horizontal strata* (T, C, K) *define invading seas and dust-dry deserts.*

lowest set of horizontal layers *(T)* exposed just above the crystalline rocks are marine sediments. They were deposited as loose sands and muds washed into an ancient sea that crept slowly up and over the beveled mountain range. A geologist looking at these strata sees the rocks as historical records of unique deposits.

Trained observers with vivid imaginations can enjoy thoughts of blue-green seas spreading across this region, leaving each rock layer as a successive record of advance and retreat. The brilliant white rocks *(C)* in the left foreground were deposited by parched desert winds, grain by grain, some 260 million years ago. The same white desert sands *(C)* are exposed on the far canyon wall as the second white strip from the top. The higher white layer *(K)* is a marine limestone about 245 million years old. This is evidence that the desert lands and all the region around them sank down and were covered by the blue-green ocean once agian. Since that time, the entire region has been lifted vertically about 2.4 kilometers (8,000 feet) above sea level. The original horizontality of the current smoothed sea floor still shows in the flat-lying attitude of the elevated strata. It is a startling sight! Think of what you are looking at, a 245-million-year-old sea floor *(K)* standing far above modern sea level.

Geologists gain both emotional and technical appreciation for the long spans of earth time by extending vicarious experiences like this, which reach far back into a remote antiquity. Almost anyone with imagination may learn to deal with earth time by constructing similar scales from their own experiences at state and national parks.

A more general scale for appreciating earth time may be constructed using the life span of the dinosaurs as a calibration unit. These familiar animals lived for about 160 million years. Their extinction occurred about 64 million years ago. This is 0.4 of a calibration unit. Abundant sea life, protected by hard, easily fossilized shells, appeared in the rock record about 600 million years ago. This is equivalent to about 3.8 calibration units. The 4.5-billion-year age of the earth represents about 28 calibration units. People have ruled the earth for a very short time. Humans in one form or another have been on earth between 5 and 10 million years. Quasi-civilized man, tool maker and master of fire, has existed for at least 200,000 years. These time spans are short compared to the 160 million years that dinosaurs ruled the earth.

Scaling by personal experience may also be applied to the measurement of distance. New York and Paris may seem far away if you have never been to those places. Similarly, space exploration seemed impossible until we landed men on the moon and instruments on other planets. Some of us who have traveled over the earth are still uncomfortably vague about places as remote as the moon. The distance 384,393 kilometers (238,854 miles) sounds more formidable than it really is. The center of the moon is only 30.2 earth diameters (one earth diameter is 12,740 kilometers, or 7,962 miles) from the center of the earth. This is about ninety-two times as far away as the direct distance between New York and San Francisco. A round-trip flight from New York to San Francisco is 1/46 of the distance to the moon.

Viewing things in scale is a very practical way to avoid becoming overawed by the mystery of size and large numbers. Travel to the moon seems much more practical when it is described in terms of earth diameters. No wonder scientists felt the trip was possible. Unfortunately there are other matters of scale that are not as easy to grasp. It has proved to be easier to visit the moon and bring back samples from this museum of creation than to explain these triumphs to the tax-paying public. Scientists have not yet learned to communicate their discoveries in any meaningful way across cultural barriers and boundaries. We have no scales for the purpose!

Viewing a World Population of 4.07 Billion People

It is hard to imagine 4.07 billion of anything, much less that many real people requiring food, clothing, shelter, medical services, and fulfillment of their dreams. This many cubic centimeters of bubble gum would require a large box 40.7 meters by 10 meters by 10

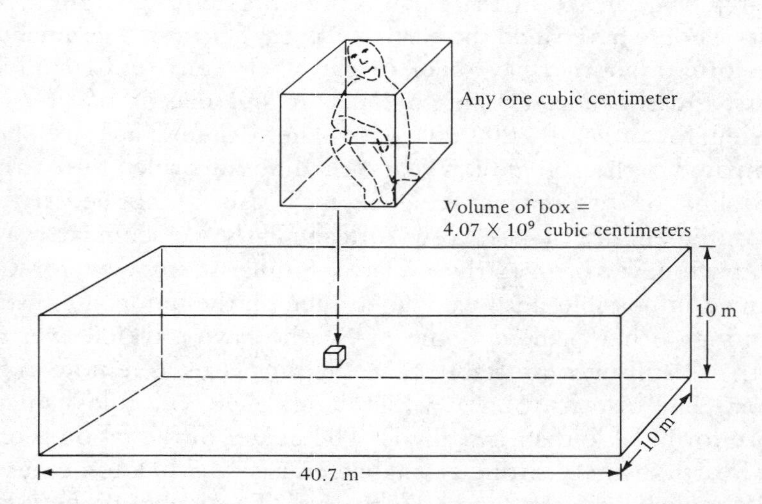

Figure 4-5 *Imagine a box 40.7 meters long, 10 meters high, and 10 meters wide filled with 4.07×10^9 cubic centimeters of bubble gum. Each cube can be thought of as representing one person in the world's population, for there are that many of us struggling to get along. Think how many of us there are clamoring for food, clothing, shelter, medical care, and fulfillment of our dreams. It is easy to see how dependent we are on what other people let us do.*

meters (133.5 feet by 32.8 feet by 32.8 feet) to hold it all. This box has a volume of 4070 meters, with each cubic meter holding a million small pieces of bubble gum.[3] A line of these boxes would be 4.07 kilometers long (2.55 miles). Fortunately, we live on a beautiful planet large enough to allow an average density of no more than 27.4 people per square kilometer (71 people per square mile). Of course, that is not the way we have spread ourselves around.

The real distribution of the world's population is shown in Figure 4-6. In some ways the map contains no surprises. We have all known since childhood that China and India abound in people, whereas Australia and northwestern Canada are almost empty lands. At the same time, there are surprises. How is it possible that vast regions are apparently underpopulated, when small places like Japan and Java are dreadfully crowded? Part of the answer is to be found in Figure 4-7, which shows the gross distribution of grain-producing areas. Note that there is an almost perfect correspondence between the two maps.

This correlation suggests that most people live where they can obtain sufficient food. The correlation also tells us that the population of the world has adjusted to the agricultural revolution that occurred about six thousand years ago, rather than to the industrial revolution of the last three centuries. There are a few exceptions to this generalization. The bulk of the U.S. and Canadian population is not located in the grain belts. We knew something of that already. At the beginning of this chapter, we learned that only about 4 percent of the working force of the United States was engaged in farming as contrasted to about 75 percent of the laboring force in China. Naturally, the Chinese would have to be clustered around the grain-growing areas to a much greater degree than Americans. Australians seem to be efficient farmers, too. That continent also has a small population living in the grain areas.

For the most part, the barren lands are either too dry, too cold, or too lacking in good soil to be productive. All these factors are geological. We see the world's population still camping out on the same sites that were settled milleniums ago as the best places to live and flourish. It would be necessary to present detailed climatic and geological data, area by area, to improve this picture. That is beyond our purpose. We merely want to see the world in first-order models. The concept of dynamic equilibrium will help us.

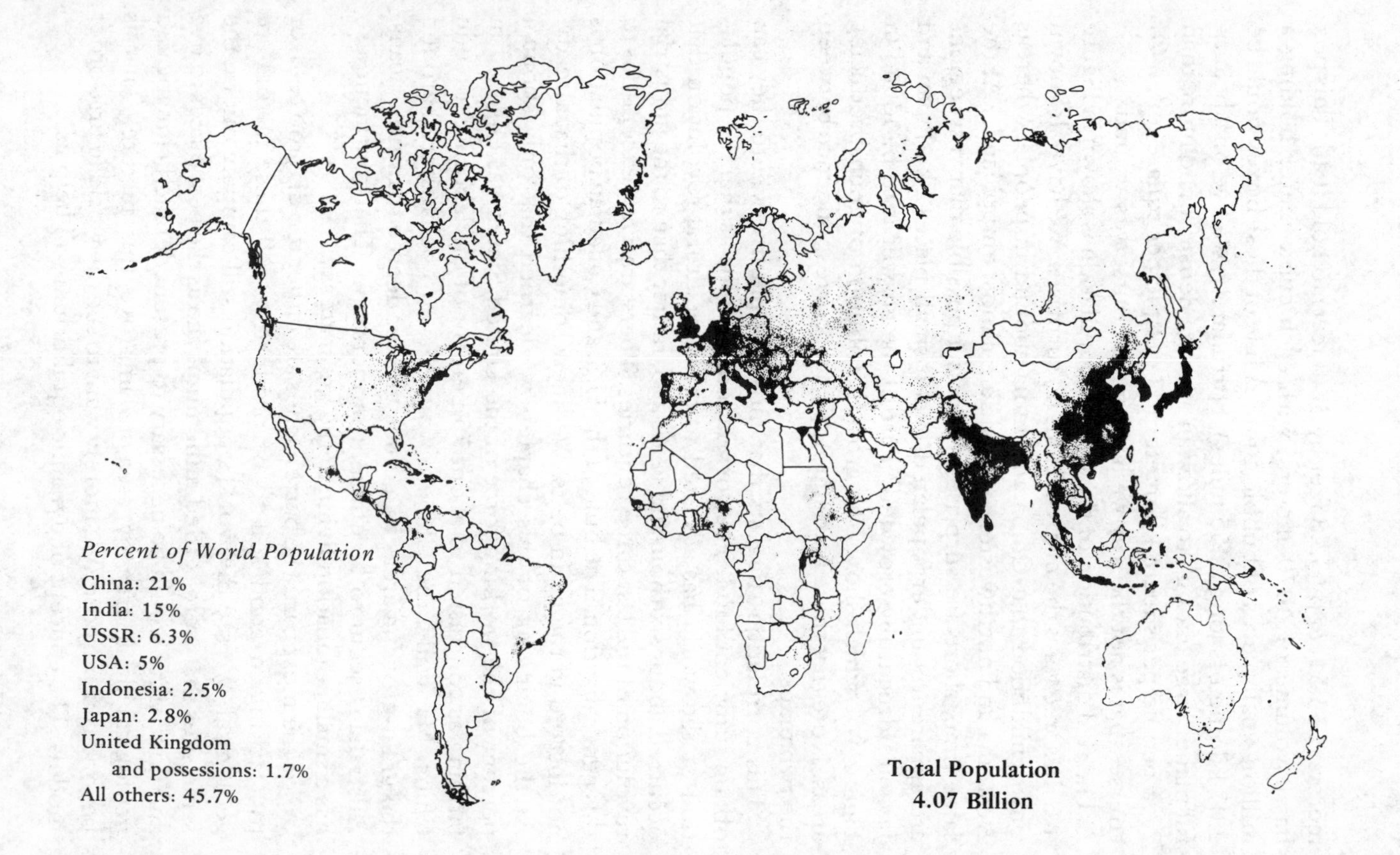
Percent of World Population
China: 21%
India: 15%
USSR: 6.3%
USA: 5%
Indonesia: 2.5%
Japan: 2.8%
United Kingdom
and possessions: 1.7%
All others: 45.7%
Total Population
4.07 Billion

Again, in dynamic equilibrium the kinds of change and rates of change are proportional to the forces involved. The Soviet Union offers a good illustration. It is a developing modern nation occupying one-sixth of the land area of the world with only 22 percent of its labor force engaged in agriculture. Nevertheless, it has problems. Most of its land lies too far north for productive farming. Low farm yield is equivalent to a low protein supply for both people and livestock. We can see this at a glance. Moscow, the capital and cultural center of the country, lies north of the 55th parallel of latitude. The same line crossing North America passes through the remote and barren country of central Labrador, the northern limit of James Bay and the southern tip of Alaska. It is cold that far north. Most Canadians live far south of the 55th parallel. Historically, the population of the Soviet Union has been engaged in a constant struggle with nature. For a long time, fish protein has been a possible solution to the threat of potential starvation. Therefore, it is not surprising to see the Soviet Union construct vast fishing fleets and an aggressive navy. The situation is predictable as a step to be taken by an industrializing nation with the resources to do it. A maritime venture is in dynamic equilibrium with Soviet environment. We may expect other nations with access to the sea, as well as an expanding industrial potential, to do the same thing.

Once again the result is predictable. A new state of change will be forced on the world when the seas have been overfished and their protein supply exhausted. That, too, is part of the scale of things.

The distribution of languages throughout the world offers another way to see the human population in scaled perspective. Languages are hard to count. There are between two and three thousand of them, plus an unmeasured number of variations and dialects. We need to see them first in historical perspective. A language is part of a culture. It is one of the most important learned

Figure 4-6 (facing page) WORLD POPULATION 1977 *Each dot on this map represents 200,000 people as they were distributed in 1972 when the world population was estimated to be 3.7 billion. Total population figures for 1977 have been added without attempting to modify the pattern of dots.*[1] *(After Laur and Guidry.*[8,9] *Timothy Michael Laur,* American Geophysical Union, *vol. 57, no. 4, pp. 189–195, 1976. Copyright by same.)*

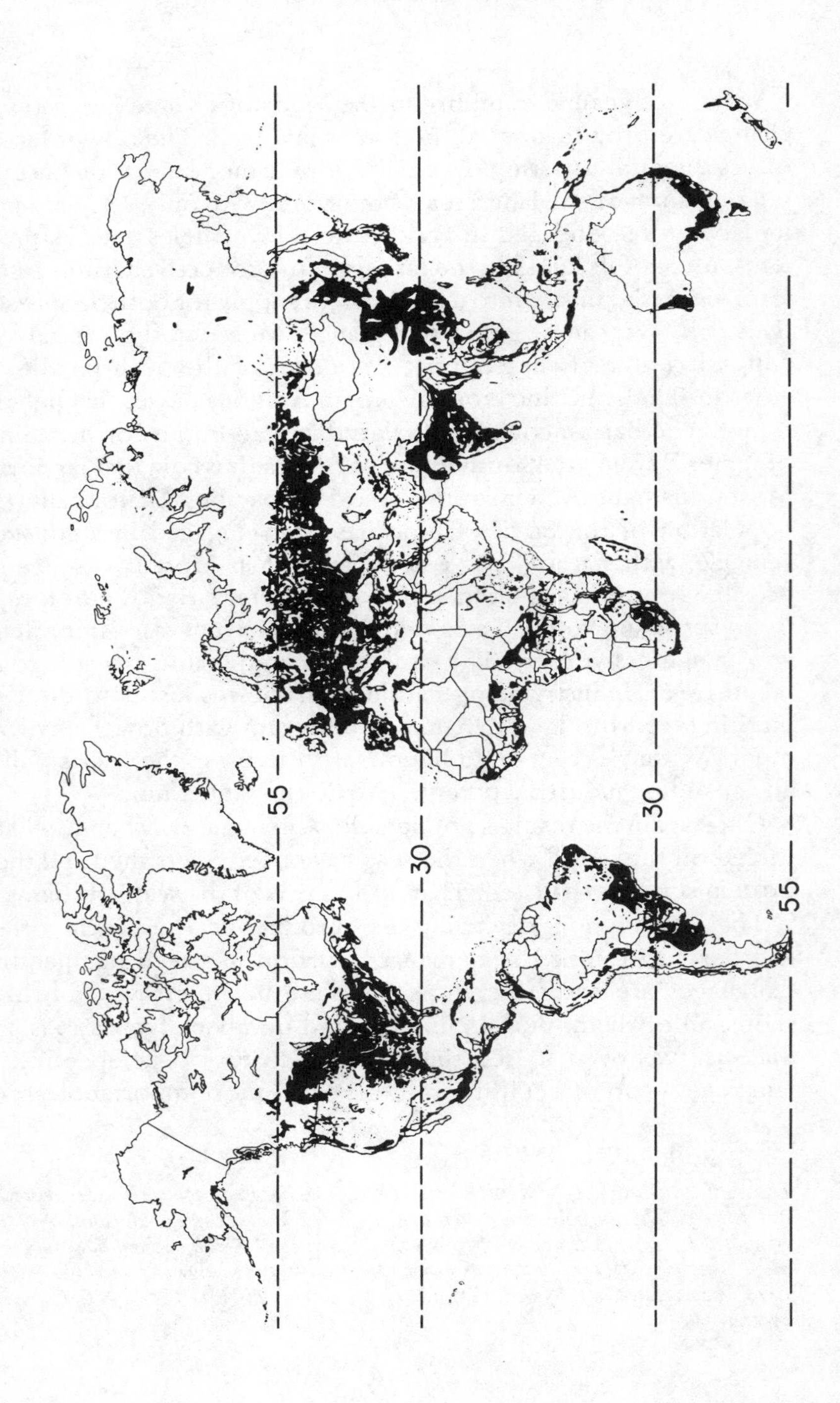
55
30
30
55

and shared ways of doing things that any population uses to maintain its identity. Irish Gaelic, Scotch Gaelic, Welsh, and Basque represent familiar ancient languages that still serve as rallying points in bitter struggles for survival.

Traditionally, the geographical limit of a language was set as the boundary of political domination by the people speaking it. Thus, a language map of the world is a picture of successful political aggression through time. Rome set the standard for the Western World, as the Latin language and Roman culture overlapped native forms. Enough uniqueness remained in each geographical area to generate new languages and new cultures following the empire's collapse. The same pattern occurred after the collapse of European colonial empires. Multinational languages and cultures are spreading around the world as a result of the political realignments established since World War II. It is a mistake to view language distribution as static.

Languages are in dynamic equilibrium. They indicate the kinds of change and rates of change that occur around us. For this reason, languages tell us a great deal about the stabilities and instabilities of our restless world. Through them, we may add a sense of scale to our own world views.

Answering Complex Questions by the Method of Successive Approximations

Scientific reasoning uses the method of successive approximations. The idea is simple enough. Plan a tentative approach, develop it step by step, add improvements as you see some inaccuracies and as new insights reveal just what the final form should be. This is a universally useful, creative method. Poems are written this way.

Figure 4-7 (facing page) GROSS DISTRIBUTION OF THE GRAIN BELTS OF THE WORLD *Notice that this pattern is nearly identical to that of the world population shown in Figure 4-6. This suggests that people live where they can get something to eat. The world population is adjusted to the agricultural revolution of six thousand years ago rather than to the industrial revolution of three hundred years ago. (After Laur.*[8] *CAED, Rep. 26, 10 pp., Iowa State University, 1966. Courtesy, Dr. Louis M. Thompson.)*

Ideas are first written down in poetic form, then words are interchanged until the most satisfactory structure emerges. Pictures are painted this way: sketch, daub, scrape off, change colors, highlight, and try again. The art of sculpture illustrates the method of successive approximations even better. Two different techniques are used. Sculptors either build up the body of a statue by adding more and more clay to the growing form or they chisel away unwanted pieces of stone to let the body emerge from its hiding place within the block.

We will demonstrate both techniques. The technique of building up larger and larger approximations will be used to answer a question about the number of grains of sand on a beach. The chipping technique will be used to estimate the number of fish in the sea.

How Many Grains of Sand Are on a Beach? The place to begin planning an answer is at a beach, where we can pick up a handful of material and see the characteristics of the population. The range of sizes, shapes, and compositions is surprising. A typical New England beach is composed of many partially rounded rocks that entrap all sizes of gravels and sands between them. Floridian beaches contain varying percentages of quartz sand mixed with broken seashells. At Miami, the ratio of quartz sand to shell fragments is about half and half. Farther south at Key West, the quartz fades out, raising the percentage of shell fragments. As much as 95 percent of the weight of beach sands from New Jersey to Georgia may contain uniformly rounded quartz sand. The rest of the material is usually a mixture of broken seashells and fine-grained, dark-colored mineral grains eroded from crystalline rocks farther inland.

We shall start by selecting a Carolina beach typical of the mid-Atlantic coast. This beach has 92 percent quartz sand and 8 percent mixed material that will not be counted. The quartz grains are reasonably uniform in character. They are not quite spherical in shape and have diameters averaging 0.3 millimeters. We first decided to determine its composition by inspecting, hand picking, and weighing random samples.

The next step is to collect three representative samples of one hundred grains each by carefully hand picking. This is an attempt to obtain samples that reflect in fair proportion all the kinds of quartz sand on the beach. Each sample is then washed to eliminate sea salt and dried at a temperature of 110°C to drive off the water. After this, the samples are weighed on a beam balance to determine their masses. Average values may then be calculated for the mass of one hundred sand grains (0.0454 grams) and the mass of one "standard, quartz sand grain" (0.000454 grams). We are now ready to deal with a larger sample.

This time we need to know the mass of a "standard volume" of sand. Once again, a random sample is selected. The procedure is crude but adequate for the answer we are after. A one-liter glass beaker is filled with wet, hard-packed sand. This, too, is carefully washed with fresh water to eliminate the salt from the sea. The sample is slowly dried in an oven at 110°C for many hours until all the water is driven off and the mass of the dry sand and mixed impurities may be determined. That value (1,756 grams) must be reduced by 8 percent to find the mass of the quartz sand without impurities (1,616 grams). The laboratory work is now completed. All that remains is to calculate the number of grains in the standard volume and expand that knowledge to the entire beach.

The calculation is done in this manner:

$$\frac{1616 \text{ gr qtz sd/liter}}{0.000454 \text{ gr/std qtz sd grain}} \cong 3.56 \times 10^{6} \text{std qtz sd grains/liter}$$

There are 1,000 liters in a cubic meter. Therefore, the number of "standard, quartz sand grains"/cubic meter is approximately 3.56×10^{9}.

Next we determine the cubic meters of sand on the beach. This is a matter of definition. There are too many variables to make any definition inclusive. The width of the beach varies greatly between high tide and low tide. The thickness of the sand is an arbitrary measurement. The shape of the beach is irregular and ever changing. For the sake of keeping things in scale, let us make a beach model 100 kilometers long, 50 meters wide, and 1 meter deep. The volume of such a feature is:

$$100 \text{ km} \times 1{,}000 \text{ m/km} \times 50 \text{ m} \times 1 \text{ m} = 5 \times 10^{6} \text{ m}^{3}$$

The total number of "standard, quartz sand grains" on such a model beach is:

$$3.56 \times 10^9 \text{ grains/m}^3 \times 5 \times 10^6 \text{ m}^3 \cong 1.78 \times 10^{16} \text{ grains}$$

That is a big number, but it is finite. Taken step by step in successive approximations, the derivation of the number seems understandable enough. Obviously, making estimates of this sort does not lie beyond human reason!

How Many Fish Are in the Sea? Estimating the number of fish in the sea is not much more difficult than estimating the number of sand grains on a beach. Our plan is simple enough. Discard the water and count the fish. We shall begin with an estimate of 1.3×10^{18} m^3 for all the sea water in all the oceans of the world. This unchallenged measurement was developed by scientists working with the United States Geological Survey.

The total volume of fish in the seas must be considerably smaller than the volume of water, but how much smaller? If we can estimate the volume of fish, we can divide it up into smaller units. Each unit will represent one fish ready to be counted.

The first step is to decide where fish live so that we may eliminate the sea-water volume that is barren of them. A map of the commercial fishing grounds of the world is some help, for it shows where fish are found in commercial quantities near enough to markets to make the business profitable.[10] More importantly, the map shows that fish are concentrated in shallow water, chiefly over continental shelves. These are the places where fish are able to find the most food. Fish eat floating plankton, swimming and bottom-dwelling invertebrates, and other fish. Most of the food chain is concentrated in shallow water areas because the food source is plant material that obtains its energy from the sun. For these reasons, we will restrict our study to the upper kilometer of the world's oceans.

Ocean basins have an average depth of 4 kilometers. The upper kilometer represents about one quarter of the volume of the ocean or 3.3×10^{17} m^3.

The second step is to decide how much of this volume is made up of fish. We need to make some trial approximations to gain perspective. One percent of a cubic meter is ten liters. If the volume of fish is as high as 1 percent of the volume of the shallow parts of the oceans, there would be ten liters of fish for every cubic meter of space. One percent is obviously too high. There would be insufficient room for an adequate supply of food and dissolved oxygen. If we lower the figure to 0.01 percent (0.0001), each cubic meter of sea water would contain 100 cm^3 of fish. That is still a very large figure, as any fisherman knows. It is true that fish are often densely concentrated in schools, but they then vacate the adjacent areas. Ten cubic centimeters of fish per cubic meter of sea water (0.00001 m^3 fish/m^3) is probably a better average figure for the portions of the sea with the greatest concentration of fish. That commercial fishing fleets experience difficulty in finding suitable concentrations indicates that even this figure is far too high for general consideration. Therefore, it is necessary to find an even lower figure for the bulk of the oceans.

Table 4-2 is a condensation of calculations made to estimate the total volume of fish in the oceans to a depth of 1 kilometer. The total volume of the oceans as estimated by the United States Geological Survey is given at the top of the table. The final figure (6.9×10^9 m^3; that is, about 7,000 km^3) represents the total volume of fish of all kinds and sizes to a depth of 1 kilometer. The next step is to estimate how many fish make up this volume. At this stage in the calculation, we may be guided by common sense and two ecological realities pointed out by Paul Colinvaux.[11]

Predators must be about ten times as large as their prey to overcome and eat them. Furthermore, predators require a great deal more caloric energy than do the individuals on which they subsist. These two ideas imply that there must be many more small fish at the bottom of the food chain than large ones at the top. The two graphs, curve 1 and curve 2, in Figure 4-8 were drawn as attempts to model the distribution of fish volumes within the population.

Curve 1 is designed to emphasize a population dominated by small fish. In this model, 80 percent of the fish have volumes of 100 cm^3 or less, while only 1 percent have volumes of 2,000 cm^3 or more. Curve 2 emphasizes larger fish. In this model, only 35 percent of the fish have volumes of 100 cm^3 or less while 5 per-

Table 4-2

CALCULATIONS USED TO ESTIMATE THE TOTAL VOLUME OF FISH OF ALL SIZES IN THE WORLD'S OCEANS TO A DEPTH OF 1 KILOMETER

Total volume of the oceans, 1.3×10^{18} m³

Volume of the oceans to a depth of 1 km, 3.3×0.10^{17} m³

Percent of the shallow oceans with highest concentrations of fish, 0.1%, (0.001)

Volume of shallow oceans with highest concentrations of fish, 3.3×10^{14} m³

Percentage of shallow oceans with intermediate concentrations of fish, 10%, (0.10)

Volume of shallow oceans with intermediate concentrations of fish, 3.3×10^{16} m³

Volume of shallow oceans with minor concentrations of fish, 3.0×10^{17} m³

Estimated highest concentrations of fish, 10 cm³ fish/m³, or 1×10^{-5} m³ fish/m³

Estimated intermediate concentrations of fish, 0.1 cm³ fish/m³, or 1×10^{-7} m³ fish/m³

Estimated minor concentrations of fish, 0.001 cm³ fish/m³, or 1×10^{-9} m³ fish/m³

Volume of fish in areas of highest concentrations, 3.3×10^{14} m³ $\times 1 \times 10^{-5}$ m³ fish/m³ $= 3.3 \times 10^{9}$ m³

Volume of fish in areas of intermediate concentrations, 3.3×10^{16} m³ $\times 1 \times 10^{-7}$ m³ fish/m³ $= 3.3 \times 10^{9}$ m³

Volume of fish in areas of minor concentrations, 3.0×10^{17} m³ $\times 1 \times 10^{-9}$ m³ fish/m³ $= 3.0 \times 10^{8}$ m³

Total volume of fish to a depth of 1 km, 6.9×10^{9} m³ (about 7,000 km³)

cent have volumes of 2,000 cm³ or more. Each curve was broken into a series of nearly straight line segments by points A_1B_1, B_1C_1, A_2B_2, B_2C_2, and so on. Limiting percentage values, such as the 5 percent separating points C_1 and D_1, were read directly from the numbers on the *x*-axis at the bottom of the graph. Median volumes for each percentage unit, such as the 300 cm³ for point $M_{C_1D_1}$ set at the 9 percent value, were read from numbers on the *y*-axis. All these values are listed beside curves 1 and 2. They are used in Table 4-3 to calculate maximum (1.81×10^{15}) and minimum (5.95×10^{13}) estimates of the numbers of fish in the sea.

None of these figures is any more reliable than the models on which they are based. Nevertheless, the figures are neither random ones nor vague beyond reason. The method was applied logically at each step. Most fish in the sea are confined to shallow waters to be

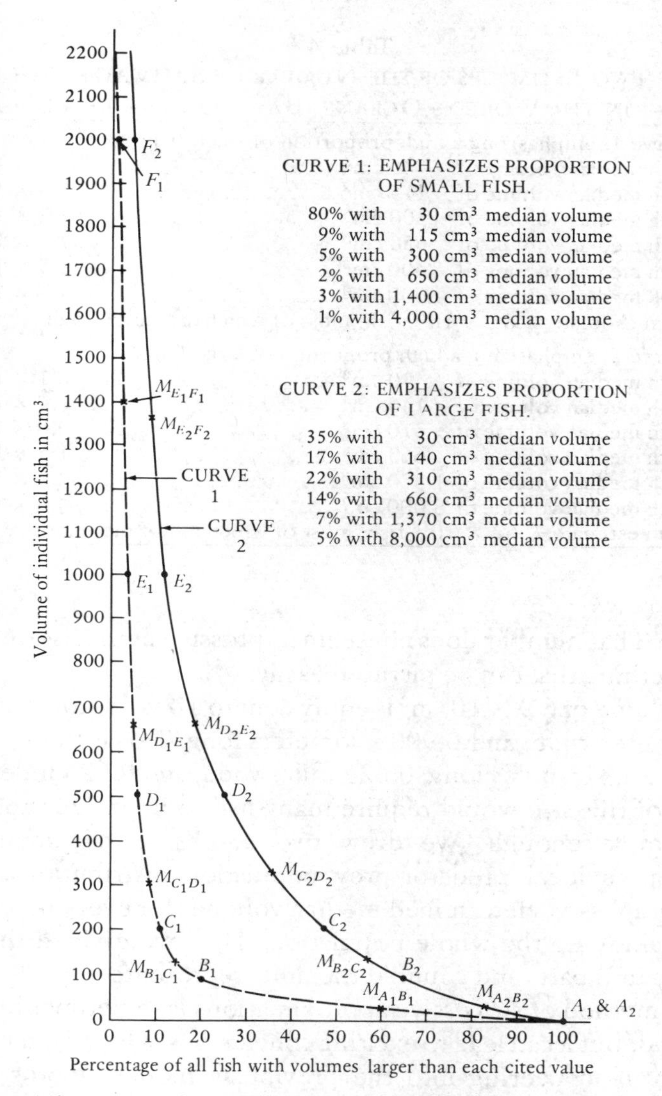

Figure 4-8 *These two graphs were drawn as unsupported estimates of the distribution of volumes among all saltwater fish. Curve 1 emphasizes small fish (80% with a median/percent volume of 30 cm³). Curve 2 emphasizes large fish (5 percent with a medium volume of 8,000 cm³). Each graph was broken into segments (A_1B_1, and so on), from which percentages of sizes and median sizes could be obtained.*

accessible to food and oxygen supplies. Different concentrations of fish were considered before adding together a total volume, 6.9 ×

Table 4-3
TWO ESTIMATES OF THE NUMBER OF SALTWATER FISH IN THE WORLD'S OCEANS (BASED ON FIGURE 4-8)

From curve 1, emphasizing a high proportion of small fish:	
80% with median volume of 30 cm^3	1.8×10^{15} fish
9% with median volume of 115 cm^3	5.4×10^{12} fish
5% with median volume of 300 cm^3	1.2×10^{12} fish
2% with median volume of 650 cm^3	2.2×10^{12} fish
3% with median volume of 1,400 cm^3	1.5×10^{11} fish
1% with median volume of 4,000 cm^3	1.7×10^{10} fish
Maximum estimate: 1.81×10^{15} fish, most of which are quite small	
From curve 2, emphasizing a high proportion of large fish:	
35% with median volume of 30 cm^3	8.1×10^{13} fish
17% with median volume of 140 cm^3	8.6×10^{12} fish
22% with median volume of 310 cm^3	4.8×10^{12} fish
14% with median volume of 660 cm^3	1.5×10^{12} fish
7% with median volume of 1,370 cm^3	3.5×10^{11} fish
5% with median volume of 8,000 cm^3	4.4×10^{10} fish
Minimum estimate: 9.62×10^{13} fish, most of which are of medium size	

10^9 m^3. That number does not seem impressive until it is converted to something that can be pictured easily.

A volume of 6.9×10^9 m^3 is equivalent to a box 1 kilometer high, 1 kilometer wide, and 6,900 kilometers long. This is the same size as a box 4,313 miles long, 0.625 miles wide, and 0.625 miles high. A box of this size would require many fish to fill it. Our approach was simple enough. We drew two graphs as fair models of Colinvaux's ideal predator-prey population distributions. From these graphs, we determined median volumes for every percentage unit making up the whole population. Then we divided the total volume into parts and counted the units represented.

The method of successive approximations is sometimes long and involved, but it can lead to exciting conclusions. There is an undeniable thrill in peering into the previously hidden depths of the oceans and counting the fish by a method as accurate as our judgment of its use. Here is an example with serious implications.

A reasonable estimate of the total catch of all the world's fisheries averaged through the early 1970s is about 68×10^6 metric tons per year. If we consider that the density of fish is approximately that of fresh water, we can convert this weight to an equivalent volume of fish. A metric ton of fresh water is equal to 1,000 kilograms with a volume of 1 m^3. Therefore, the volume of the world's yearly fish catch is equal to 6.8×10^7 m^3. This is about 1 percent of the total volume of fish calculated to be in the world's oceans. (6.8×10^7 m^3 fish caught per year divided by 6.9×10^9 m^3 total fish in oceans = $0.0098 \cong 1$) If these figures are reliable, we have already begun to overfish our oceans, destroying the breeding herd on which so many people depend for protein.

4-4 The Dark-Night-Sky Paradox

We have been learning how to view things in proper scale. This means gaining perspective by finding ways to see nature in big units without losing too much detail. Our illustrations have been confined to large populations contained within reasonably well-defined limits. Let's expand this perspective to include the whole universe. Our definition of universe includes everything there is: the earth, the sun, stars, galaxies, black holes, and all other matter and energy wherever it may be. New questions are inevitable. How many of these objects are included in the universe? Where are they in relation to one another? What are they like? Can we count them as we did the sands of the beach and the fish of the sea? Do we dare to approach the study of the universe in a quantitative way?

The answer to that question is a qualified yes. We can begin to think quantitatively if we first assume some particular but as yet unproven shape for the universe. Imagine that the universe has a spherical shape that can be described in the three rectangular dimensions of length, width, and height. This assumption is not defendable, but it will do for the purpose of our discussion to show the value of viewing things quantitatively in scale. Kepler will give us encouragement.

Johannes Kepler (1571–1630) was a German astronomer who grasped the importance of trying to think quantitatively. He felt there was no alternative. "Just as the ears are made for sound and the eyes for colors, when the mind leaves the realm of quantity it wanders in darkness and doubt."[12]

One of the questions that perplexed Kepler was spawned by the earlier astronomical studies of Thomas Digges. In 1576, Digges had made a sweeping generalization that utterly destroyed the sun-centered universe, which had been the legacy of Copernicus. Digges's idea made good sense. He saw that every star was a sun in its own right, "farr excellinge our sonne in both quantitye and qualitye."[13] We still agree with his conclusion. If every star is a sun and if our sun is nothing but a common star, there is no reason to imagine that we occupy the center of the universe. Digges saw no reason to imagine that the universe had any boundaries at all. Kepler was not so sure. The existence of a dark night sky bothered him.

Kepler's doubt about an infinite universe was a brilliant perception. He thought that an infinite universe filled with an infinite number of very bright stars would have to be equally bright in all directions. There could be no night in such a universe, because billions of stars as bright as the sun would represent an enclosing set of overlapping light sources. Under these conditions, Kepler saw no way to explain the dark night sky with which we are all so familiar. Here, then, was a paradox. On the one hand, there was good reason to imagine that the universe could extend infinitely in all directions. On the other hand, he felt that the universe must be limited in size because the night sky is dark.

A large forest is a good analogy to show Kepler's reasoning. Someone standing in the forest could look in any direction and see nothing more than a wall of tree trunks. The full size of such a forest would remain unknown to a stationary observer because tree trunks would block all lines of sight.

The dark-night-sky paraxdox gained a new dimension between 1720 and 1744. Two astronomers, Edmund Halley and J. P. Loys de Cheseaux, decided that the power of light diminishes in some way between distant sources and the earth. If this were true, the night sky could appear dark rather than bright because light from distant sources would be lost. The paradox became important again after Albert A. Michelson (1852–1931) had successfully measured the constant velocity of light.

Astronomical research in the twentieth century has revealed other startling things about the universe. We know that there are literally billions of galaxies, each containing some 100 billion individual stars. Each star radiates light in our direction without destroying the darkness of the night sky. Every star is bright. There are just not enough of them to fill the sky. That, too, is odd and requires an explanation.

The answer to this, as well as to the dark-sky paradox, may be expressed in many ways.[13] All the explanations add a new dimension — time — to the method of successive approximations. We know that stars, like our sun, are born, radiate their energy, and die in a period of about 10^{10} years. During that time, their light traveling at the velocity of 3.0×10^5 km/sec (1.86×10^5 mi/sec) would cover a maximum distance of 9.5×10^{22} km (5.9×10^{22} mi). That is as far away as we could hope to be able to see an actual object that is still producing light. We would know nothing of more distant stars that are too young for their light to have reached us. There is also the possibility that very faint undetected light from extremely distant and long extinct sources is passing us today. If so, it is lost in the scattered background radiation that makes up the dark night sky.

Our visible universe is relatively small. We know nothing of its more distant reaches. Imagine the night sky eons from now as it begins to light up with rays from beyond our present horizons and loses light from stars that have exhausted their fuel and died. Would it not look very different from the universe we now behold?

Science is humbling. The more we learn about nature, the less certain we are of our abilities to see things in proper scale. Sometimes it is best to follow Walt Whitman out into the night.

WHEN I HEARD THE LEARN'D ASTRONOMER

When I heard the learn'd astronomer,
When the proofs, the figures, were ranged in columns before me,
When I was shown the charts and diagrams, to add, divide, and measure them,
When I sitting heard the astronomer where he lectured with much applause in the lecture-room
How soon unaccountable I became tired and sick,
Till rising and gliding out I wander'd off by myself,
In the mystical moist night-air, and from time to time,
Look'd up in perfect silence at the stars.[14]

This is the true dark-night-sky paradox.

Annotated References

1. *Goode's World Atlas,* 15th ed., ed. Edward B. Espenshade, Jr. and assoc. ed. Joel L. Morrison (Chicago: Rand McNally, 1978).
   These figures were taken from graphs accompanying a world map of the predominant economies, pp. 26–27, and from population tables, p. 230.
2. J. Oliver Thomson, *History of Ancient Geography* (New York: Biblo and Tannen, 1965).
   This book is a tremendous compilation of information on the geographical thinkers of the ancient world. Many of them made contributions to our present world view and yet have been forgotten, like Pytheas, or slighted, like Eratosthenes. Scholars will find pp. 143–151 and 157–168 rewarding reading.
3. Alexander Moszkowski, *Conversations with Einstein* (New York: Horizon Press, 1940).
   The quotation about the blind bug appears in the Introduction, p. xviii.
4. Armen Alchian and William R. Allen, *Exchange and Production: Competition, Coordination and Control,* 2nd ed. (Belmont, Calif.: Wadsworth, 1977).
   The six postulates on which the science of economics is based are found on pp. 24–27.
5. Steve H. Sanders and John Pilley, "Effects of Alcohol on Timing Behavior in Rats," *Quarterly Journal of Studies on Alcohol,* 34 (1973), 367–373.
6. United Nations, *Demographic Yearbook 1976,* 28th issue (New York: Department of Economic and Social Affairs, Statistical Office, 1977).
   Information on Samoa is found on pp. 940–941.
7. U.S. Bureau of the Census, *Statistical Abstract of the U.S. 1976,* 97th ed. (Washington: U.S. Department of Commerce, Bureau of the Census, 1977).
8. Timothy Michael Laur, "The World Food Problem and the Role of Climate," *Transactions of the American Geophysical Union,* 57 (1976), 189–195.
9. N. P. Guidry, *A Graphic Summary of World Agriculture,* misc. publ. 705 (Washington: U.S. Department of Agriculture, Government Printing Office, 1964).
10. *Oxford Economic Atlas of the World,* 4th ed., ed. D. B. Jones (London: Oxford University Press, 1973).

The maps on pp. 26–27 give locations and production figures for commercial fish of all sizes.

11. Paul Colinvaux, *Why Big Fierce Animals Are Rare* (Princeton, N.J.: Princeton University Press, 1978).
    The title of this book is taken from the third chapter. The subtitle, *An Ecologist's Perspective,* applies directly to the problem of defining our approach. Dr. Colinvaux is a brilliant analyst, with a surname that appears to offer an amusing contrast. *Colin* is definitely of either Irish or Welsh origin. *Vaux* is French for "lazy, slow-witted fellow." It suggests that some long-forgotten ancestor was one of the "wild-geese" who went to France to join the Irish Brigade, which fought England in the eighteenth century. If so, he was probably renamed by the French soliders.
12. H. Butterfield, *The Origins of Modern Science, 1300 to 1800 A.D.* (New York: Macmillan, 1953).
    Kepler's interesting opinion about quantitative thinking is on p. 67.
13. E. R. Harrison, "The Dark Night Sky Paradox" *American Journal of Physics,* 45 (1977), 119–124.
    Harrison's method of settling the paradox is to show that the universe will never be saturated with light. As new stars are born others die out, keeping the total amount of light constant. The king is dead. Long live the king. Fine, but we still have just one king, never two.
14. Walt Whitman, *Leaves of Grass* (New York: David McKay, 1900).
    Whitman has much to say to those who will tarry and listen.

FIVE

PLAYING FOR HIGH STAKES: THE MOTIVATION OF SCIENTISTS

Francis Bacon . . . spoke of making observations,
but omitted the vital factor
of judgment about what to observe
and what to pay attention to.

Richard P. Feynman[1]

5-1 Enlarging a Limited Understanding of Science

We have already seen how scientific revolutions take place when a crisis of contradiction forces scientists to see nature in a new way. It is now time to learn how that knowledge can enlarge our own limited understanding of science. It is time to learn how to use what we already know.

We begin with the two models of the fabric of science in Figure 2-9. One shows the fabric as seen by scientists. The other shows the student's view of it. These models are similar in one important way. Each exhibits a boundary that separates understanding from ignorance. Scientific progress at all levels requires discovering the true nature of nature hidden beyond that boundary of ignorance, wherever it may be. Students face the task of learning things that other people already know. Creative scientists discover things that no one has ever known before. Both groups can use the same learning technique. We will study it by examining the familiar adventure of Robinson Crusoe and the footprint in the sand. Writing in 1719, Defoe's purpose was much broader than simply telling a story. *Robinson Crusoe* contains an analytical view of social mores. We can pull out one analysis and use it to illustrate the steps in our own technique of learning.

Step 1: Recognize New Data

The scene opens as Robinson Crusoe notices a human footprint pressed in beach sand well above the limit of high tide. He had never seen such a thing in the entire fifteen years since his shipwreck and had every reason to believe that no other human was on his island.

> It happened one day, about noon, going towards my boat, I was exceedingly surprised with the print of a man's naked foot on the shore, which was very plain to be seen in the sand. I stood like one thunderstruck, or as if I had seen an apparition: I listened, I looked round me, but I could hear nothing, nor see anything; I went up to a rising ground, to look farther; I went up the shore and down the shore,

> but it was all one; I could see no other impression but that one. I went to it again to see if there were any more, and to observe if it might not be my fancy; but there was no room for that, for there was exactly the print of a foot, toes, heel, and every part of a foot: how it came thither, I knew not, nor could I in the least imagine; but, after innumerable fluttering thoughts, like a man perfectly confused and out of myself, I came home to my fortification . . .[2]

Crusoe spent several days and nights barricaded within his fortress home fearing that strangers on the island might kill him. At last he had an encouraging thought. Perhaps, this was one of his own footprints left in the sand after landing his boat on the shore. He decided to go back to the beach and measure the print against the size of his own foot. Now, we have established the need for the second step (please excuse the pun).

Step 2: Verify the Data To Be Sure There Has Been No Error in the Observation

> . . . but when I came to the place, first it appeared evidently to me, that when I laid up my boat, I could not possibly be on shore anywhere thereabout: secondly, when I came to measure the mark with my own foot, I found my foot not so large by a great deal.[2]

The data have been recognized, observed, and verified. But the humanness of the observer may play tricks with his or her scientific detachment. Feynman states it nicely when he speaks of the "vital factor of judgment about what to observe and what to pay attention to."[1]

Step 3: Accept the New Data as a Valid Part of the Now-Enlarged Fabric of Knowledge

This means that the boundary between knowledge and ignorance has been moved to include a new enlarged view of nature. The crisis of contradiction has appeared and has destroyed any former complacency. The observer's view of nature will never be as provincial as it was before. As Robinson Crusoe summed it up, "Both these things filled my head with new imaginations."[2]

Accepting startling new information is not easy. Emotions may become involved. Scientists have often found this stage of a revolution very taxing. Rather than accept the new information and the crisis of contradiction that comes with it, individuals may simply tune out the information and continue to function as if nothing has happened. A classic example was the reaction to Alfred Wegener's observations on continental drift. The present position of the continents cannot explain why there are extensive 250-million-year-old glacial deposits in India so close to the equator. This fact alone should have alerted the entire geological community to continental movements through time. For many of us, particularly those living in the northern hemisphere, the idea was unbelievable simply because it was too grand. Our refusal to think seriously about it was self-defense.

Progress cannot be made that way. It is necessary to accept new data and think about it. This is much easier to do if we just realize what has happened. New data takes us into an area that was once one of total ignorance. It offers us a place to stand, an island of certainty from which we may look back at the familiar pattern of the fabric of science. Just one gap of ignorance separates the new data from the old. We must bridge that gap with the most logical explanations possible. Only then will we feel emotionally comfortable and intellectually satisfied.

Step 4: Use the New Data as an Intellectual Foundation from Which To Construct an Improved Logical Model of Nature

> I went home again, filled with the belief that some man or men had been on shore there; or, in short, that the island was inhabited . . .[2]

This last step is purely intellectual. The crisis of contradiction ends because a new model of nature is made with a newly defendable least-astonishment explanation for the way things are. Let us look at Robinson Crusoe's thought patterns more closely.

The footprint was just a small piece of circumstantial evidence on an otherwise featureless beach surrounding a small island that had been known to be uninhabited for fifteen years. Nevertheless, the footprint undeniably belonged to another human being, perhaps a cannibal. Crusoe's judgment would not permit him to overlook that fact. The stakes were too high! Therefore, he was forced by the principle of least astonishment to conclude "that the island was inhabited."[2] Robinson Crusoe had been forced to move intellectually from his old model of nature to a new one. Here we have an important generalization. *Cognitive learning occurs when a person is forced by logic to accept a new intellectual position because the stakes are too high to follow any other course of action.* This point is certainly worth examining.

The Idea of Playing for High Stakes as Motivation for Scientists

It would be difficult to imagine a physical scientist whose contributions are held in higher regard than those of Albert Einstein (1879–1955). The games he played with his creative talents were certainly for high stakes. He was so successful at it that it also may be hard to imagine he could have ever doubted how to use his life. Nevertheless, the young Einstein, like many other children, wondered and searched for a satisfying way to use his instrument (brain of the ages). The story is best told in his own words, as he looked back on his sixty-seven years of striving and searching.

> Even when I was a fairly precocious young man the nothingness of the hopes and strivings which chases most men restlessly through life came to my consciousness with considerable vitality. Moreover, I soon discovered the cruelty of that chase, which in those years was much more carefully covered up by hypocrisy and glittering words than is the case today. By the mere existence of his stomach everyone was condemned to participate in that chase. Moreover, it was possible to satisfy the stomach by such participation, but not the man in so far as he is a thinking and feeling being. As the first way out there was religion, which is implanted into every child by way of the traditional education-machine. Thus I came — despite the fact that I was the son of entirely

Figure 5-1 *Albert Einstein (1879–1955) was about twelve years old when he discoverd: "Out yonder there was this huge world . . . which stands before us like a great, eternal riddle. . . . The contemplation of this world beckoned like a liberation . . ."*[3] *(Courtesy of the Library of Congress)*

irreligious (Jewish) parents — to a deep religiosity, which however, found an abrupt ending at the age of 12. Through the reading of popular scientific books I soon reached the conviction that much in the stories of the Bible could not be true. The consequence was a positively fanatic (orgy of) free-thinking coupled with the impression that youth is intentionally being deceived by the state through lies; it was a crushing impression. Suspicion against every kind of authority grew out of this experience, a skeptical attitude towards the convictions which were alive in any specific social environment — an attitude which has never again left me, even though later on, because of a better insight into the causal connections, it lost some of its original poignancy.

It is quite clear to me that the religious paradise of youth, which was thus lost, was a first attempt to free myself from the chains of the "merely-personal," from an existence which is dominated by wishes, hopes and primitive feelings. *Out yonder there was this huge world, which exists independently of us human beings and which stands before us like a great, eternal riddle, at least partially accessible to our inspection and thinking. The contemplation of this world beckoned like a liberation, and I soon noticed that many a man whom I had learned to esteem and to admire had found inner freedom and security in devoted occupation with it. The mental grasp of this extra personal world within the frame of the given possibilities swam as highest aim half consciously and half unconsciously before my mind's eye. Similarly motivated men of the present and of the past, as well as the insights which they had achieved, were the friends which could not be lost. The road to this paradise was not as comfortable and alluring as the road to the religious paradise; but it proved itself as trustworthy, and I have never regretted having chosen it.*

. . . In a man of my type the turning-point of the development lies in the fact that gradually the major interest disengages itself to a far reaching degree from the momentary and the merely personal and turn toward the striving for a mental grasp of things. Looked at from this point of view the above schematic remarks contain as much truth as can be uttered in such brevity. [Italics mine.][3]

High stakes are measured by the beholder. Einstein saw high stakes in the opportunity to liberate himself for a life devoted to "striving for a mental grasp of things."[3] His motivation as a scientist began quite early when he determined to respond to the opportunity to play this game. A firm definition of motivation may be based on this game.

Motivation for any activity is simply the determination to respond to an opportunity to play for high stakes. Creative scientists share a single motivation: the thrill of discovery. They glory in it. Scientific interests usually show up during childhood, but it is inaccurate to imply that all scientists specialize so early. A good deal more exposure is usually necessary to test their interests and to define sufficiently exciting high stakes before they establish career motivations. Here is an example from the life of James S. Coleman, one of the most productive sociologists in the United States.

Coleman's research made him a public figure early in his career. He is the principal author of such classic studies as *Union Democracy: The Internal Politics of the International Typographical Union, Social Climates in High Schools, The Adolescent Society: The Social Life of the Teen-Ager and Its Impact on Education, Models of Change and Response, Uncertainty and Medical Innovation: A Diffusion Study.* His most famous research was done for the United States Office of Education and entitled *Quality of Educational Opportunity.* Although other authors contributed, this study has become known as the "Coleman report."

This report discloses the effects of public educational policies prior to its publication in 1966. Segregated schools are a primary issue. Technically sound sociological research is extremely important because the results may be used to formulate new public policy. When Dr. Coleman writes about the ideal of creating a "science of social phenomena," he is pointing out the high stakes that are involved. Here is his reflection on the critical moment in his life when he saw a future for himself helping to develop "a science of social phenomena":

> A prospective sociologist ordinarily aspires to becoming one only relatively late, after his habits of mind have been formed. Thus the critical juncture for most sociologists was not when they began to use evidence systematically to test ideas, but when they first saw that this can be done in areas of human behavior.
>
> The time of the latter revelation to me was during college, several years after I had determined to become a research engineer or physical scientist. In a course in political science my professor described a new method for measuring the attitudes of persons through their responses to a series of questions. I had been thinking about the voting patterns of U.S. Senators on foreign policy, before and after the Second World War. It seemed to me that some Senators expressed consistent isolationism or interventionism in these two periods, while others took positions that seemed to be dependent on whether the external threat was seen to be Hitler and Germany, or Stalin and Russia. Sure enough, this method, applied to votes of the same Senators in 1941 and 1948 showed these

different patterns. Each Senator could be characterized, not by a long list of his votes, but far more concisely, by his position on a scale at these two times, and thus by the difference or similarity of his positions.

A new world suddenly opened up to me: it was possible to study human behavior with methods of observation and analysis that could give systematic increments of knowledge. A science of social phenomena was feasible and I might be able to help develop it.[4]

What Creative Scientists Must Do

Creative scientists must live simultaneously in two different intellectual worlds. Day-to-day research is a systematic hunt that takes place methodically after a decision has been made to follow one approach in the hope that it will produce results. At this level, scientists employ intellectual routines in which dependable technique is an important requirement. Even the words that describe this standard research have a calm, ordered quality. The second intellectual world of creative scientists is a startling contrast to the first. It is disordered, full of surprises and unimagined opportunities. In this world, the creative brain searches for clues to larger systems that may be hidden in the data. A kaleidoscopic mixture of intuition, imagination, vision, daring, and judgment replace dependable techniques as the most desired qualities. Rewards depend on the ability to guess correctly. Good fortune often presents critical facts at the most unanticipated moments. The creative scientists are the ones who recognize these opportunities and act on them.

Oddly enough, good fortune in science does not seem to be a matter of chance. It seems to be produced by scientists who have been thinking seriously about particular problems. These are the lucky few who gain critical insights from stray happenings in which they suddenly see special meanings. The famous story about Sir Isaac Newton and the falling apple is an excellent example.

Any attempt to reconstruct exactly what Newton thought in a quiet moment at age 23 or 24 when he saw the apple fall must be fictionalized somewhat. We simply do not know all the details. Contemporary sources close to Newton tell part of the story, but omit saying that Newton had prior knowledge of the laws of motion. We will return to this point after reading two early accounts. The first one was written by Newton's friend William Stukeley.

> On 15 April 1726 I paid a visit to Sir Isaac at his lodgings in Orbels buildings in Kensington . . . After dinner, the weather being warm, we went into the garden and drank thea, under the shade of some apple-trees, only he and myself. Amidst other discourse, he told me, he was just in the same situation, as when formerly, the notion of gravitation came into his mind. It was occasion'd by the fall of an apple as he sat in a contemplative mood. Why should that apple always descend perpendicularly to the ground, thought he to himself. Why should it not go sideways or upwards, but constantly to the earths centre? Assuredly, the reason is, that the earth draws it. There must be a drawing power in matter: and the sum of the drawing power in the matter of the earth must be in the earth's centre, not in any side of the earth. Therefore dos this apple fall perpendicularly, or towards the centre. If matter thus draws matter, it must be in proportion of its quantity. Therefore the apple draws the earth, as well as the earth draws the apple. That there is a power, like that we here call gravity, which extends its self thro' the universe.
>
> And thus by degrees he began to apply this property of gravitation to the motion of the earth and of the heavenly bodys, to consider their distances, their magnitudes and their periodical revolutions; to find out, *that this property conjointly with a progressive motion impressed on them at the beginning,* perfectly solved their circular courses; kept the planets from *falling upon one another, or dropping all together into one centre;* and thus he unfolded the Universe. This was the birth of those amazing discoverys, whereby he built philosophy on a solid foundation, to the astonishment of all Europe. [Italics mine.][5]

A more technical understanding of Newton's thoughts is given by Henry Pemberton, editor of the third edition (1726) of the *Principia* (that is, Newton's *Philosophiae Naturalis Principia Mathematica, Mathematical Principles of Natural Philosophy,* first published in 1687).

> The first thoughts, which gave rise to his *Principia,* he had, when he retired from Cambridge in 1666 on account of the plague. As he sat alone in a garden, he fell into a speculation on the power of gravity: that as this power is not found sensibly diminished at the remotest distance from the center of the earth, to which we can rise, neither at the tops of the loftiest buildings, nor even the summits of the highest mountains; it appeared to him reasonable to conclude, that this power must extend much farther than is usually thought; why not as high as the moon, said he to himself? and if so, her motion must be influenced by it; perhaps she is retained in her orbit thereby.[6]

This was a bold step. The young Isaac Newton had daringly related by common cause two simultaneous events that had occurred in sight of one another, but were 384,000 kilometers (239,000 miles) apart. One event was the falling apple that he saw dropping straight down from the tree toward the center of the earth. The other event was the falling moon that he also saw dropping straight down toward the center of the earth. This relationship may have seemed so simple to him that he never felt the necessity to tell anyone how he did it. Perhaps that is just as well, for it leaves us an opportunity to be creative in our own right and at least to share vicariously the thrill of discovery with him.

We need to make one reasonable assumption, Newton could not have made the generalization that the pull of gravity influenced the orbital path of the moon, unless he had already discovered what we now call his first law of motion. *A body at rest will remain at rest unless an unopposed force (a push or a pull) acts upon it. A body in uniform motion at constant speed will continue to move in a straight line unless compelled to change direction by an unopposed force.*

It would have been strange indeed for Newton to be ignorant of these facts as late as 1666, because Galileo had already drawn attention to them in his book, *Discourses Concerning Two New Sciences,* published in Holland by Elzevier in 1638. Furthermore, Newton considered the laws of motion so basic to understanding nature that he stated them as axioms as the beginning of his *Principia.*

Figure 5-2 shows both the earth and the moon as we might see them from a position in space far above the North Pole. Looking

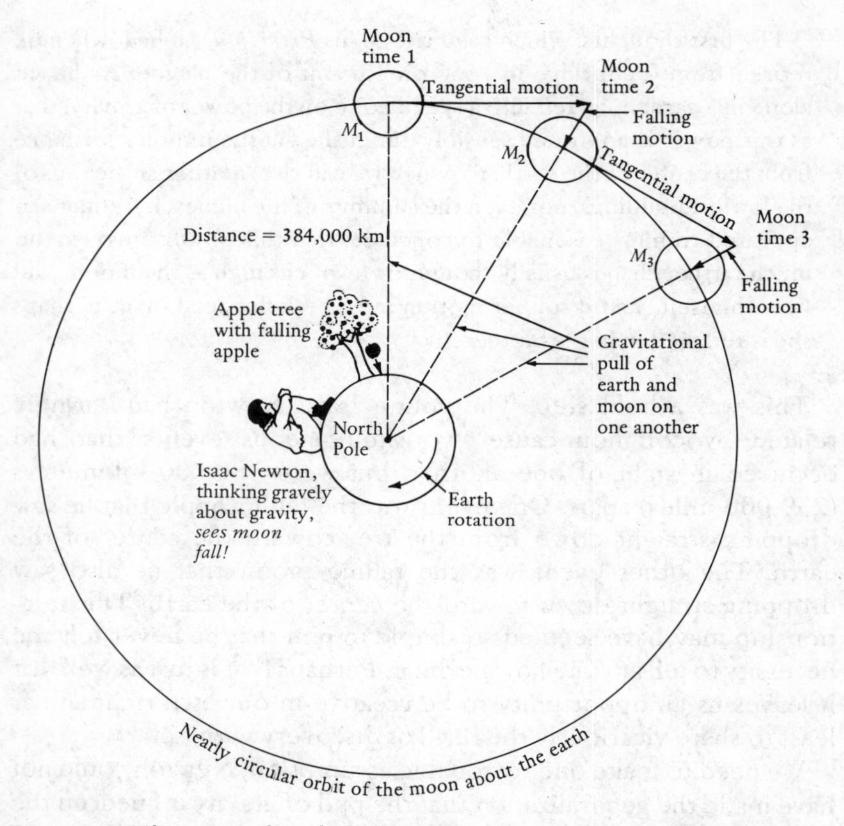

Figure 5-2 *This picture shows both the earth and the moon as we might see them from a position in space far above the North Pole. As we look down, we can see Isaac Newton watching the apple fall at the same time he sees the moon fall down toward the center of the earth. The familiar orbital motion of the moon around the earth may be thought of as taking place in small, simultaneous increments, over and down. Imagine being the first person to realize this!*

down, we can see Isaac Newton watching the apple fall at the same time that he sees the moon fall down toward the earth. The moon is a body in uniform motion that circles the earth once every 27.32 days. We know by the first law of motion that the moon should move in a straight line, if a tangential push dating from an event in Creation is the only force acting upon it. However, the moon orbits the earth in a nearly circular path. Therefore, another force, the gravity between the earth and the moon, must constantly pull the moon back toward the center of the earth. The familiar orbital motion of the moon around the earth may be thought of as taking place in a series of small, simultaneous increments, over and down. Imagine being the first person to realize this!

Newton's own reaction to his discovery is interesting. We have glimpsed it in William Stukeley's account.[5] Voltaire (1694–1778), who obtained his information from Madame Conduit, Newton's niece, gives a more complete story.[7] As soon as Newton saw the apple fall from the tree, he began a series of mental processes.

> What is this force . . . he asked himself. . . . It acts on all bodies in proportion to their masses, and not their surface area; it will act on the fruit falling in the orchard even if lifted 3,000 toises [one toise equals 1.95 kilometers, 1.21 miles], even if lifted 10,000. If that is so, then the force must act from the location of the moon right to the center of the earth. If so, this power (whatever it is) could be the same as that which pulls the planets toward the Sun and which attracts the satellites of Jupiter toward the center of Jupiter in their motion around Jupiter.[7]

Once Newton had assured himself that gravity acts at great distances from the earth, the next step was obvious. He needed to determine the strength of the gravitational field between bodies like the earth and the moon or between the sun and Jupiter. It took him almost twenty years to develop a defendable proof of what we now call his law of gravity. The stumbling block was an uncertainty that the mass of each body could be considered to be acting as a single value placed at the geometric center of its volume. The same point is found in William Stukeley's account of the scene in the garden, when Newton asked himself, "Why should that apple always descend perpendicularly to the ground? Why should it not go sideways or upwards, but constantly towards the earth's centre?"[5] Newton invented differential and integral calculus as tools with which he was finally able to solve the problem, resulting in his well-known law of gravity. Its chief elements are: the amount of force between two attracting bodies is directly proportional to the product of their masses and is inversely proportional to the square of the distance between them.

We have been pointing out what creative scientists must do. They must function simultaneously in two different intellectual worlds.

Day-to-day research is a systematic hunt that takes place methodically after a decision has been made to follow one approach in the hope that it will produce results. The second intellectual world is one in which creative scientists search for clues to larger systems that may be hidden in the data. Sir Isaac Newton had his own answer to the problem of success as a creative scientist.

> A visitor once asked Newton how he discovered the laws of the world. "By thinking about it ceaselessly," he answered. "This is the secret of all great discoveries: genius in the sciences depends only on the intensity and duration of thought that a man can muster.[7]

What Creative Teachers and Creative Students Must Do

Scientists have their own social culture ruled by their learned and shared ways of doing things. Creative scientists, creative teachers, and creative students represent specialized subcultures within the larger group. Belonging is simply a matter of doing. Individuals earn knighthood in the subculture by performing feats of comprehension. Their transition from beginners to acknowledged discoverers is not mysterious like the transition from a larva to a cocoon-wrapped pupa to a butterfly. It occurs where thoughts are intense, are of long duration, and are creative.

Creativity is a process of giving birth. The Latin root *creare* means "to produce or to make." A much older Indo-European root *ker* means "to cause to grow." This idea gets right to the heart of the matter, particularly when linked with the suffixes *ive* and *ivity* from the Latin *ivus,* "having a tendency to cause to grow." Intensity and duration of thought furnish the power for creativity, but a triggering mechanism is necessary to release the tendency to cause to grow. Newton's view of the apple falling straight down toward the center of the earth was more than just seeing an apple fall. It was a triggering mechanism, a view that gave birth to an idea and caused the idea to grow. Creative teachers and creative students must also make ideas grow. In this way they become members of the clan, Isaac Newtons, Junior Grade.

There are some common misconceptions about the arts of teaching and learning that inhibit creative behavior. A teacher should never stand between the students and the discipline, concealing important ideas behind a dominating screen of regulations and methodology. Obedient students functioning in intellectual lockstep are unlikely to look up and see the moon fall. Furthermore, students should never expect or be allowed to let their teachers do their thinking for them. You cannot learn to swim if the teacher does all the swimming.

Creative education is a partnership. Teachers know the well-marked trails by which students may reach the edge of the wilderness that beckons to be explored. Undergraduates have an advantage that many more advanced students and professionals have lost. Every view is a new experience to most undergraduates. The thrill is much like that of traveling for the first time to a new land where everyone is strange and every object exciting.

Intellectual explorations are just as animating. Imagine looking up the roots for the word *creative* for the first time and discovering that the sound *ker* means to cause to grow. Reading further, we find that *ker* is also the root sound for cereal and for Ceres, the Roman goddess of agriculture. An entirely new chain of thought is suddenly revealed. A time sequence has just been introduced. These words are latecomers to our language. There is no reason to assume they predate the agricultural revolution of some six thousand years ago. Is it possible that the entire Indo-European idea of creatively designed weapons, utensils, and clothing are much older? Did the arts of making things predate the generalization we now call creativity? Are such clues to the processes of socialization dependably revealed in language roots? Instead of answers, anyone daring enough to let imagination ramble will find more penetrating questions, ready to be savored.

Curriculum planners coerce students' thoughts to flow through a full range of academic disciplines. This range is a tremendous help in learning to be creative, for it places seemingly unrelated ideas next to one another. The sequential reasoning of mathematics is as well displayed in linking words into sentences, sentences into paragraphs, and paragraphs into themes as it is in solving a calculus problem. Students learn about life from each other and from contrasting images in literature. Human hopes, follies, and triumphs are laid bare in religion, history, and art. Variations in human culture analyzed in the social sciences have parallels in the life sciences, literature, and language. The physical sciences reveal a world that often dominates what living populations can do under conditions found in different environments. Nations are suddenly revealed as captives of place rather than just as rulers of people. Undergraduates find themselves to be wanderers in this wonderland, freely licensed to weave ideas together to make their own unique insights. Four years of their lives are spent in these pursuits. The creative students among them often develop a passion for the practice of thinking, "That's strange. I wonder what it means."

Creative teachers of science share with students a need to see themselves as productive scientists rather than just fellow travelers. Routine activities get in the way of following Newton's example of thinking with intensity and duration. Nevertheless, the teaching routine offers a compensating opportunity to be creative. Every teacher must prepare for every lecture and classroom experience. These must be thought out daily. "Here's the material. How shall I approach it?" Teachers may take this opportunity to learn something new and to have the joy of talking about it. These are high stakes! Great teachers are not hard to identify. They will always leave the students with something to think about.

5-2 Finding the Ability to Place Seemingly Small Things in Larger Contexts and to Make Significant Generalizations

It is one thing to follow the brilliant reasoning of someone like Sir Isaac Newton and quite another to do similar things without guidance. Is this really something that less talented people can hope to do? Can relatively ordinary students and teachers hope to make their own significant scientific generalizations? The best answer may be a deferred one: wait and see. Proof of genius is the accomplished fact. Perhaps, we must wait until after we have produced scientific generalizations before we know we can make them. Scientific education in America is partly at fault for our indecision.

We may be operating under an unchallenged system that exists for the convenience of the faculty. Students of science (unlike students in the humanities who must write many original term papers) are not expected to produce new generalizations of their own until they write dissertations for advanced degrees. Before then, almost all tests and examinations are metered against the clock. Students are expected to come to them prepared to spill out all they know in a hurry. There is no opportunity for the occurrence of a stray event to trigger a creative muse. There is not even an opportunity to think intensely over enough time to permit someone to derive a new equation and solve a surprise problem. Students are brought along in batches, a roomful at a time. Under these conditions, it is not surprising for them to be unaware of their own creative abilities. To be unaware is not the same thing as being incompetent. Students may only need encouragement to work in ways more professional than those demanded by the educational system.

Encouragement is exaggerated communication designed to give heart. The word comes to us from the French *cour* and the older Latin *core,* both meaning "heart." There is nothing scientific about encouragement. It is an emotional experience. Encouragement brings hope and this emotion lies far beyond the sterile limitations of science.

A psychologist attempting to study the relationship between moodiness and indecision would be hard pressed to design a definitive set of scientific experiments that carry the impact of *Hamlet.* Shakespeare had no Ph.D. in clinical psychology, but he did know how to assume and to use literary license as he examined major problems reflected in the lives of individuals like Lear, Othello, and Richard III. Our approach shares this ambition. We will examine anecdotes about common insights that people of all ages have had as they discovered something that was to them shockingly new. In this way, we hope to show that the ability to make significant discoveries is not so much a rare quality as it is an unnecessarily neglected capacity in all of us. It is waiting, ready to be used if called upon.

The first two anecdotes deal with two important generalizations that most people probably make at an early age. One is the distinction between self and everything else. The other is recognition that the individual is consciously thinking. Note that in each case a visual observation offers new data that lie just beyond the limit of prior understanding. The brain then makes the logical ties necessary to link the new and old data together as part of a more harmonious world.

The earliest incident in which I can remember having been aware of a rational pattern of thought occurred when I was two years old. I was riding in a car with my mother and one of her friends on an air base in Africa. An airplane was preparing to land and the two ladies asked if I could see it. Instead of an airplane I saw what I now realize was an irregularity in the fluid of the eye which anyone can see either when looking through a microscope or when viewing a clear blue sky. I had noticed this phenomenon before, but the request to look for an airplane placed it in a new perspective. I knew that what I saw was not an airplane, but it had the attributes of one. As I saw it, it was up in the sky and it appeared to be flying, since it moved when I moved my eye. In short for the perceiver, the essence of this visual phenomenon coincided with my knowledge of airplanes. Knowing that the ladies were expecting me to see the airplane landing and realizing that what I had seen would do for one, I replied, "Yes, I see an airplane."[8]

I was around the age of three when one morning I was aware that I was behind the curtains of the living room of my home. I realized the curtains were moving and that my brother was beside me. From there I ran to the kitchen and realized that my mother was cooking breakfast. I spoke to her. The awareness of having other things around me started at that moment.[9]

The next two anecdotes reveal a growing awareness of social reponsibility. There is a two-stage pattern in these accounts. The first stage is characterized by an externally imposed authoritative viewpoint. The second stage is marked by a deep personal realization of the responsibility that these children share in the well-being of others. Their world views have been expanded to include people who are hidden from sight.

When I was about four years old I had a small bow that came equipped with suction-cup tipped arrows. After a while I discovered that I could tighten the string by wrapping it around one end of the bow. In the process I also found out that the arrows could be made to travel farther, if the suction cups were removed. I showed my discovery to my mother, omitting the part about the suction cups, of course! I did point out that the arrows would probably go over the house, if I let them. She cautioned me not to do that.

Well of course, some time later I did shoot one over the house. I can still see that arrow going up and over the roof until it disappeared from sight. I raced around to the back yard to retrieve it and see how far it traveled. There was my mother hanging clothes. I was scared, then horrified as the vision of the arrow piercing the top of her head and killing her popped into my mind. I never did find that arrow, but what I did find was vastly more serious and valuable — especially at that age.[10]

I was between three and four years old when a handsome young stranger from the big city visited at the house two doors from ours. This was an event. I can remember him sitting on the porch one sunny summer day effusing a warm brown glow from the most beautifully tailored suit I had ever seen. The attraction drew me to him and up onto the throne of his knee. Once we were acquainted, an easy step between a child and a happy young man, he offered me a penny for a kiss. It seemed like a good bargain to me. I kissed him slyly on the cheek and ran home in high glee to tell my mother.

She was furious. . . . At him and at me! Suddenly the sun was gone behind a mysterious black cloud of the hate only grownups know. Years passed before I had any idea why my Eden had collapsed and why I had been driven from the garden, apple in hand. I have since come to realize that the innocence of children and of happy handsome young people is rarely taken at face value.[11]

The following four anecdotes show us something about the value of combining guidance with personal experience. The first is a remarkably creative situation ending with a total loss and a frustrated child who had no idea that anything more could be done. The second is a case of partial loss, because the faculty did not recognize the importance of a student's observation and react to it. The third and fourth anecdotes represent total successes. Instructors were present and took the opportunity to make memorable events of seemingly minor things. The third one is particularly interesting because not only did the child, Linus Pauling, have the instrument to forge and use his new generalization, but we are familiar enough with his career to appreciate that fact.

When I was between four and five years old, my family lived in Washington, D.C. My father took us for a drive one night in our 1922 Buick touring car. Stoplights had not yet been installed in the downtown area and the police were directing traffic. I asked my mother why they were wearing white gloves. She told me that this was to make their hands more visible. The next day, I had an exciting idea. If white was good, red should be immeasureably better. I would create light! There is an unforgettable thrill in being that age and in knowing for certain that you are thinking. I was so keyed up that I could barely perform the experiment I had designed to test my idea.

I tore a small piece of shiny red cardboard from a Buster Brown shoe box in which I kept some toys. Even then in the excitement of the moment I was practical. The torn piece was only about the size of a penny. I needed the box and didn't want to destroy it. Red cardboard in hand, I went into my closet and closed the door to darken the space. A bar of light was still visible at the crack along the bottom of the door. This worried me at first and I was preparing to cover the crack when I realized with horror that it didn't make any difference. I couldn't see the red cardboard anyway. The excitement drained away as I was forçed to admit that I hadn't invented a light source after all.

The emotional depression that followed is still vividly clear to me. I gave up on red and made a half-hearted, resigned effort to test white with a piece of writing paper. That too was a failure. I couldn't see the paper clearly with the closet door closed. That was that. I gave up and went outside to play with my friends. My mother was very bright. I realized now that I should have told the story to her and found out what went wrong. She knew enough physics to help me understand the difference between new light and reflected light.[12]

I recall a realization that occurred to me during my first year in graduate school and which had a profound influence of my career. My major professor in psychopathology felt students needed to experience life before textbooks would make much sense to them. He assigned me to a somewhat regressed schizophrenic who had considerable difficulty in communicating with me. Her thoughts would jump about almost randomly from one context area to another. There was no coherence and no logical pattern. I was unable to understand her mental processes.

Several weeks later, I had an unusually bizarre dream and wakened with the jumbled thoughts still vividly in mind. Suddenly it dawned on me that my schizophrenic patient's mind was working while awake just as my own mind had done when I was asleep.

I told several of my faculty about it but found them too busy to be interested and too familiar with this sort of thing to be surprised. Fortunately, that didn't stop me thinking about this. I discovered that she could be perfectly coherent while reading out loud to me and was fully capable of following someone else's thoughts. Her difficulty was only revealed when she was forced to depend on her own sense of continuity and order. The lesson has been very valuable in helping me communicate with schizophrenics. I have an empathy for them for I sense that I have been there myself.[13]

When I was about six years old we lived in Condon, a small town in eastern Oregon. One day I was trying to sharpen a pencil with my knife. I was unsuccessful until a cowboy said that he would show me how. He pointed out that there was a relationship between the angle at which I held the blade of the knife and the slice of wood that was removed from the pencil. I was told to hold the knife firmly at the same angle as I made the cuts around the end of the pencil. I was so impressed by the fact that one could attack a problem in a logical way that I have remembered this episode all my life.[14]

During my last year in high school a wonderful thing happened to me in a course run by General Electric Company as preparation for a job in their labs. The following sequence of events occurred in a matter of seconds. I thought of a question to ask the teacher. I raised my hand and he recognized me. Then before I could state the question I discovered that I had figured out the answer for myself. I said, "Oh! I know the answer." The teacher followed up by wanting to know the question as well as the answer. Then we had a quick complimentary exchange, two minds in excited agreement.

It was important to me because it was the first time that I realized my own potential. This teacher was someone to whom I had looked up, because he was the most intelligent person I knew. His compliment meant that he accepted me on his level for having the capacity to think for myself.[11]

Classification is a problem in science. Establishing criteria for both kind and category is never more than a matter of judgment as to what to include and what to exclude at each level. Judgment depends on experience and experience is an ever-changing factor. There is little wonder that lumpers and splitters are rarely satisfied with each other's work. Here are two anecdotes that reflect the problem of establishing identities under conditions of limited experience.

Dennis O'Leary and I were great friends when we were both five years old and lived in Greenwich Village. Neither of us could read but we did know that letters made words and stood for sounds. We would try to spell out signs on wagons and things like that. One day we discovered a manhole cover with the letters D.P.W. on it. We had never heard of the Department of Public Works, but were sure that the letters had something to do with a sewer. Dennis solved it quite satisfactorily. D.P.W. stood for Dirty Pee Water.[15]

By the time I was six or seven I was allowed to leave our yard and go down to a private place I loved beside a small creek in the woods. A school of shiny black polywogs fascinated me with their antics. For some reason I failed to visit them for a time. When I next saw them I was astounded to find that they had changed completely. They had grown legs! My brother who was 12 years older than I explained that this was perfectly natural and encouraged me to watch them turn into frogs. For years after that I would go back to my creek and see what was new and what was repeating its cycle. So many things were there, the cowslips, the salamanders, pussy willows, horse tails and fantastic dragonflies. Everything changes, nothing remains the same. That idea has been a dominating factor in the way I view the job of living.[11]

Scientists of all persuasions, students, teachers, and researchers share the same common need: the wise use of time and resources. Everything is an investment made in expectation of future rewards. The last anecdote shows good planning as essential for success.

When I was 13 years old my father taught me to drive a car. After a few sessions on a country road I thought I knew all about it and was ready to travel. When Dad came home from work I prevailed upon him to let me solo around the block. He agreed and I went out confidently, keys in hand.

For some reason I had to fight the car all the way, just to make it go. It shuddered and groaned and would move forward without stalling only if I kept it in low gear. When I finally got home the engine was hot, the radiator steaming, the brakes were smoking, and I thought the car was about to burst into flames. It was only then that this crestfallen kid realized that he had driven the car with the hand brake solidly engaged.

I didn't admit the experience to my father, but it taught me to think for myself a little more. It got me into the habit that I have today of challenging all situations with the question, *"Do we know what we are doing?*[16]

All of us have had similar experiences from which we have drawn generalizations that have shaped our lives. Thinking about them as proof that we have been able to place seemingly small things in larger context may encourage us to approach science in the same way. One further illustration may help you make the transition from the everyday world to science.

Examine this limerick, keeping in mind that it has been presented without explanation. Think of it as a stray event that has caught your attention.

There once was a student named Hart.
Teachers never thought he was smart.
'Cause he wasn't a whiz
At their memory biz,
Scoffed Hart to his friends,
They gimme the bends,
Memory won't break a horse to a cart.

It is time to test yourself. How deep are the meanings and generalizations that you can draw from this silly verse? If you have any success in passing from the trivial to the profound, you may be stepping into the future. Perhaps, you have even considered a career in science.

5-3 What Does a Career in Science Involve?

Life is about doing things. It takes time to do things, so life is also about the use of time. We have only one chance to use each moment before it slips away. The value of a life is a measure of what an individual thought was worth doing moment by moment. A *career* is the name we give to a consistent use of working time between adolescence and retirement, senility, or death. Careers are worth the trouble it takes to plan for them, because the lifetime stakes are so high.

It may be wise to decide on the career rewards that appeal to you most before making a career decision. Table 5-1 contains a representative list of the sorts of things that many people think of as important career rewards. Look it over as a guide to identifying your own goals. Each item that is boldfaced is thought to be an

Table 5-1
CAREER REWARDS

Love ideas	Childhood ambition	**Enthusiasm for topic**
Money	Advancement	Power to wield
Feel useful	Be happy	**Have responsibility**
Travel	Be healthy	**Gain fame and glory**
Adventure	**Opportunity**	**Help people**
Romance	**Communication**	**Intellectual freedom**
Fun	Easy work	Grow emotionally
Risk	Security	Be part of a team
Satisfaction	No pressure	Stay in home town
Challenge	**Always learning**	**Enjoy thinking**
Recognition	Have title	**Play with fancy toys**
Elite group	Free time	Easy to qualify for
Be creative	**High standards**	**Devotion to a cause**
Short hours	Leader	Good location, climate
Good family life	Long vacations	Paid for being shrewd
Integrity valued	Good retirement	Expense account
Company car	**Have vision**	Collective bargaining
Enjoy precision	**Offers immortality**	**Professional identity**

integral part of most scientific careers. A comparison will suggest how well suited you are to becoming a professional scientist.

Science is an international subculture, separate in many ways from the rest of civilization. It has its own learned and shared ways of doing things. Some of these career rewards must be seen in the context of the subculture to be understood. Other items are self-evident and need no explanation. For example, it is obvious that scientists must be dedicated to the love of ideas, feel enthusiasm for their topics, share professional identities, and enjoy thinking. On the other hand, the uniqueness of being a scientist is shown best in the way the fraternity looks at some of the other rewards. We shall examine these in detail, starting with a most important reward.

MONEY

Scientists are well paid for their work. They share the same pay scale enjoyed by other highly paid professionals. University and college teachers make less than equivalent scientists employed by industry, but enjoy compensating benefits. Teachers have more freedom than industrial scientists do. Summer vacations and other long breaks during the year offer ample time for earning money by consulting, writing textbooks, and doing research. An energetic teacher may be much better paid than an industrial counterpart. This is only part of the story.

Although scientists demand and receive money for their work, money is not their sole reward. The best of them are such enthusiasts that their experiences and ambitions drive them on far more effectively than would money alone. Score keeping in science is measured by intellectual productivity.

FEEL USEFUL, HELP PEOPLE, DEVOTION TO A CAUSE

Most of us want to feel useful and help others. Scientists are no exception, but they are devoted to a very specialized cause: the discovery of the nature of nature through basic research. No one else will find the cures for cancers or concern themselves with black holes in space. Cancer research is understandable in the eyes of the general public. It is obviously practical, but black holes may seem to be silly things on which to spend money. The same thing might have been said about the researches of Copernicus. All he did was discover that the earth is not the center of the universe. Although everyone may not consider this fact important, it does have a profound effect on some religious views. Scientists are devoted to the discovery of truth and to expressing it openly for anyone to use.

OFFERS IMMORTALITY, RECOGNITION, FAME, AND GLORY

It is hard to find a person who does not have some sense of immortality. It may be reflected in religious opinion, hope for the success of one's children, or in the pride of personal achievement. Scientists share all these values and add another: the pride of participating in a great constructive adventure. Reputations are gained by making heroic intellectual leaps. Only a few scientists are sufficiently gifted to reach the highest glories, but almost any productive person may achieve acceptance as a thinker. Glory is like wine. The taste is heady enough to make addicts of gentle people.

TRAVEL, ADVENTURE, RISK, ROMANCE, FUN, COMMUNICATION, ALWAYS LEARNING

These rewards exist in many forms. Field sciences like geology, biology, oceanography, and anthropology often provide adventurous careers with a great deal of unavoidable traveling. Sometimes traveling can be fun. When it interferes with a normal family life, traveling loses its adventurous side and becomes a career cost rather than a reward. There are many other sources of fun, romance, and adventure. Searching for elusive bits of truth in a laboratory may not require physical risk, but many scientists find it just as exciting as climbing mountains. Adventure is in the eye of the participant. Discovering how Sir Isaac Newton saw the apple fall straight toward the center of the earth was an adventure we have shared with him from the comfort of our armchairs. One universal characteristic of scientists in their appreciation for the intellectual adventure of learning new things day by day.

Figure 5-3 *The glory drive is real. It shows up in many ways as people attempt to become outstanding achievers in everything from tending gardens to politics. Scientists are no exception.*

Communication is a major part of intellectual adventuring. Voltaire communicated to us across the centuries. We do not happen to live in his time, but the story of Newton and the apple remains fresh and exciting. There are two reasons for this. Voltaire chose his subject well. He understood the importance of an almost trivial happening. The second reason is that we can appreciate how Newton's genius was great enough to lift him out of his own times and into the intellectual world of the future. This is what genius does. It carries the progress of civilization into the future. That is adventurous traveling indeed!

Elite Group, Professional Identity, Responsibility, High Standards, Integrity Valued, Enjoy Precision

Science is uncompromising. Its practitioners consider themselves to be an elite group that is cohesive because they share common ideals. The work of discovery will succeed if it is done with sufficient precision and reported faithfully. Nothing must be falsified or concealed. Responsibility for intellectual honesty lies with each individual researcher. If anyone breaks the code, excommunication is automatic. Their work is no longer trusted. Scientists are proud of their record of intellectual integrity. Few other professions have done as well.

Risk, Intellectual Freedom, Challenge, Enjoy Thinking, Vision, Be Creative

There is no way to be creative without taking the risk of failure. Vision is the principal tool of creative people. They can see beyond ordinary limits. One anecdote will make the point. Professor Grope once gave a geology class the problem of making a drawing to show how a crystalline igneous rock was composed of minerals joined together as a rigid solid. One young man turned in this picture:

```
   P
   Y
QUARTZ
   O   M
   B   I
   O   C
 FELDSPAR
   E
```

It was a risky thing to do. Rocks and minerals are not made of letters shared in this way. At the same time, the picture is a visionary model of the way minerals could share electrostatic bonding between the atoms in different crystals. Old Grope wanted to find out how the student had managed to think of minerals in this way. The best approach seemed to be the direct one: just ask! The next time Grope met his class he said, "Where is Mr. Brady?" An arm waved at the back of the room. "Tell me, how did you do it?" The class sat silently, unaware of the meaning of the question.

Mr. Brady answered very simply from the back corner of the room. "I knew that our grades would depend on answers to this problem and another one. Since I knew the answer to the second problem, I felt I could afford the risk of turning in that kind of a drawing instead of one you expected."

People who enjoy thinking and being creative also enjoy the intellectual freedom that these activities produce. Scientists do not have to stand in line and wait for recognition through seniority. Their work is not judged right or wrong by some democratic vote of approval. Their work simply has to be good enough to prove its own case. This is risky business. It takes daring to be great. There is a challenge, but think of the thrill of success!

LOVE IDEAS, ENTHUSIASM FOR TOPIC, PLAY WITH FANCY TOYS, OPPORTUNITY

Scientists live on and for ideas. One of the interesting things about their devotion to ideas is the apparent restriction scientists place on the range of things with which they concern themselves. Many of them insist on devoting their energies to their own ideas and on tuning out those of other people. Odd as it seems, it is true. Psychologists do not seem to care about detailed analyses of the chemistry of granite. Geologists seem immune to the physical chemistry of carbon. Physicists find it difficult to get very excited about monocotyledons. The arachnids are of little concern to sociologists. Political scientists specializing in state and local government are totally unconcerned with gravity waves. What is important is that each kind of scientist must be an enthusiast about his or her own topic. Enthusiasm is an emotional reflection of intellectual ferment. It is a good sign that a brain is working.

The opportunity to play with fancy toys excites scientists beyond measure. Nuclear physicists like to tamper with the big machines that spill out subatomic particles for analysis. Fully automated electron microprobes are beyond the dreams of some workers, and represent the minimum that will draw other scientists to new campuses. Hardware is expensive to buy and to operate. The day has passed when scientists could watch apples fall from trees and proclaim universal laws. Seeing is no longer done with the naked eye alone. Instruments do the work. Expensive and fancy toys have become the necessities of the present. Scientists find that their futures as productive researchers are linked with their ability to use these toys. Fancy toys represent opportunities. Imagine what space-age hardware has meant to astronomers, who have been partially blinded by our turbulent, dusty atmosphere for milleniums.

SATISFACTION

This reward is given automatically for outstanding achievement. It is the sum of everything else.

Identification and Training of Scientists

Embryonic scientists identify themselves by their interests and daring. They know they have to go to college and so they show up ready to take some favorite subject in their freshman year. Those attending major universities are often lost for a year in the shuffle of large classes. Potential scientists are more readily seen in small colleges because they have more opportunity to get to know some faculty member well enough to share their dreams. Most of them have the brash confidence to last through the second year and into the third. At this level, thinkers begin to be distinguished from memorizers.

Advanced work in science is concentrated in the last two years of an undergraduate curriculum. By this time, students have completed most of the courses designed to broaden them as human beings and they are beginning to feel a sense of professional identity. The third and fourth years are exciting times; they should produce an explosive development of interests in technical things. If this interest is lacking, something may be wrong with the student's career choice. Fortunately, the end of college marks a distinct break. Career choices may be adjusted at this time.

An education in science at the bachelor-degree level has many uses. Teaching in primary or secondary school systems is one possibility. Graduates may find jobs in research and quality-control laboratories throughout industry. Their initial tasks tend to be repetitive and restricted to the routines of making standardized measurements. Industrial scientists with only a bachelor's degree often drift into administrative positions where their lack of technical preparation is less important than their abilities to assume responsibility. One of the most enticing opportunities open to science graduates with a bachelor's degree is a shift toward engineering positions. Still another choice appeals to a large percentage of science graduates. They are free to become generalists, rather than specialists. This is an open-ended choice that begins with a complete break away from science. Insurance, accounting, selling, government work, and agriculture are possibilities. The list could go on and on, even leading to specialization in other directions like the ministry, medicine, and law. In short, an undergraduate science education is not so specialized that it limits further career choices. Graduate work is more specialized and requires greater commitment.

We need to consider two graduate programs, one leading to a master of science degree, the other to a doctor of philosophy degree. The master's degree is essentially an industrial and high school teaching degree. The work is broad based rather than specialized. It is designed to cover the principal fields of physics, chemistry, economics, and so on, in one or two years. A doctorate is a research degree based on highly specialized work done over three to five years. The student becomes an apprentice under the direction of a small committee of concerned professors. These two degrees are not ordered in series. It is not necessary to have a master's degree to qualify for a doctorate.

Degree requirements vary from school to school but usually cluster around a common set of hurdles. A master's candidate will take between twenty-four and thirty academic hours of advanced course work. A twenty-four-hour program usually requires additional original research plus a formal report called a thesis. Comprehensive written and oral examinations are customary. A doctoral program begins with an exhaustive oral examination, used to determine what the candidate knows and does not know. This is followed by the assignment of 50 to 85 academic hours of advanced course work designed to cover the fields and fill the gaps. Original research followed by a scholarly dissertation is an absolute necessity. Some residence time on the campus is usually required to ensure that the degree candidate has an opportunity to live with and be a part of the scientific subculture. Rigorous oral and written comprehensive examinations are much-dreaded hurdles that come at the end of the work. Failure at this level can be a crushing experience that must be risked to enjoy the thrilling experience of success. (Ad astra cum grano salis.)

Life in graduate school usually develops into strange love-hate relationships between the students and the discipline. Love of the discipline pulls them through a traumatic struggle with their time schedule and the degree requirements. One of the great shocks that sweeps over most beginners is finding themselves full-time students. Very few students have had this experience as undergraduates. Campus activities distracted much of their energies and subordinated the academic work. This is not true in graduate school, where students eat, sleep, and dream about their disciplines. Hero worship is common at the graduate level. Students begin to meet the great names of science, people who will talk to them as colleagues. It is interesting to watch graduate students at conventions. They read the name tags of everyone in the halls and on the elevators, looking for King Arthur, Queen Guinevere, Lancelot, and Galahad. It is strong wine to stand beside a Nobel prize winner and wonder what is flashing through such a brain at that moment. These are the years of acceleration.

Students are advised to work under the best masters that will tolerate them. They are encouraged to work on the most important problems they can hope to solve and with the most advanced techniques. Some graduate students view their dissertations as annoying hurdles and the doctorate as a union card for a life in the profession. This is an error that, if pursued for long, will lead to a modest career at best. A far better attitude is to want to find a problem that is big enough to win instant respect. With this thesis as a base, a student can move readily into more challenging and more rewarding post-doctoral professional work. Career management is something that should be thought out very carefully.

Many scientists function best when surrounded by brilliant and energetic colleagues who spur one another along. Oddly enough, science, the cooperative venture, thrives on competition and competitive ideas. A scientific group capable of sustaining productivity is probably composed of about ten to twenty members who are in their most vital years. A large university or a major research team is required to furnish a group as big as this. The fancy toys are found in these places also. Obviously, the competition to get a start in a great science department or laboratory is not to be underestimated. The best credentials are evidences of past productivity and undeniable potential to become an international figure. Recommendations from great scientists are a help, but less important than the tangible proof of outstanding published research.

Graduate students should realize that these final years of training are really the first years of their professional lives. Creative scientific productivity is expected of them. The stakes are high in graduate school, particularly in one where the professors are at work reshaping our future understanding of the nature of nature. This is the graduate school to attend.

What Is It Like Out There?

Undergraduates are often trapped in the role of following a beaten path as part of a large group of fellow travelers. They run with the pack and try to please everyone. Graduate school changes all that. Dissertation research is not done by running after the pack. It requires striking out in daring new directions and pushing into the unknown at considerable risk. The undergraduate way of doing things lacks daring. It must be stiffened with the individualism so characteristic of the cat family. Creative scientists are trailblazers. They are part of Rudyard Kipling's, "The Lost Legion."

> There's a legion that never was 'listed,
> That carries no colours or crest,
> But, split in a thousand detachments,
> Is breaking the way for the rest[17]

If the idea of "breaking the way for the rest" appeals to you, you may want to consider a career in the great game of science. There are functions for an enormous range of talents.

Science, the Optimistic Ongoing Enterprise of Students, Faculty, Technicians, Administrators, Gamblers, Knights, and Superstars

Therefore doth heaven divide
The state of man in divers functions,
Setting endeavor in continual motion;
To which is fixed, as an aim or butt,
Obedience: for so work the honey-bees,
Creatures that by a rule in nature teach
The act of order to a peopled kingdom.
They have a king and officers of sorts;
Where some, like magistrates, correct at home,
Others, like merchants, venture trade abroad,
Others, like soldiers, armed in their stings,
Make boot upon the summer's velvet buds,
Which pillage they with merry march bring home
To the tent-royal of their emperor;
Who, busied in his majesty, surveys
The singing masons building roofs of gold,
The civil citizens kneading up the honey
The poor mechanic porters crowding in
Their heavy burdens at his narrow gate,
The sad-eyed justice, with his surly hum,
Delivering o'er to executors pale
The lazy yawning drone. I this infer,
That many things, having full reference
To one consent, may work contrariously:
As many arrows, loosed several ways,
Come to one mark; as many ways meet in one town;
As many fresh streams meet in one salt sea;
As many lines close in the dial's centre;
So may a thousand actions, once afoot,
End in one purpose, and be all well borne
Without defeat.

Shakespeare, *Henry V* (I:ii)

That indeed is the pattern of science.
Unnumbered actions, designed to end in one purpose:
Progressive discovery of the nature of nature.
A pattern well born, once afoot, on-going, well borne,
Without defeat.

Now recruiting
For the next campaign!
Bring forward the pipers. . . .
The adventure begins here.

Annotated References

1. Richard P. Feynman, "What is Science," *The Physics Teacher* 7 (1969), 313–320.
The quotation about Francis Bacon is part of an interesting aside in which William Harvey is quoted as describing Bacon's philosophy of science as "the science that a lord-chancellor would do."
2. Daniel Defoe, "The Life and Strange Adventures of Robinson Crusoe," *The Works of Daniel Defoe* (New York: Thomas Y. Crowell, 1903).
This selection was taken from pp. 172–178.
3. Albert Einstein, "Autobiographical Notes," in *Albert Einstein: Philosopher-Scientist,* Paul Arthur Schlipp, Editor (La Salle, Ill., Open Court Publishing Co., 1970), p. 781. Permission to quote granted by editor and publisher.
This is no ordinary autobiography. It is a mine of dazzling gems. This selection is taken from the English text on pp. 3, 5, and 7.
4. James S. Coleman, personal communication.
A number of prominent scientists were asked to contribute anecdotes from their own development to show how they discovered that doing science is a matter of reasoning. Many of them indicated difficulty in trying to recall real rather than contrived incidents. Dr. Coleman's response was one of the most lucid of the positive responses.
5. William Stukely, *Memoirs of Sir Isaac Newton's Life: 1752, Being some account of his family and chiefly of the junior part of his life,* ed., A. Hastings White (London: Taylor and Francis, 1936), p. 85.
We have quoted from p. 19.
6. H. Pemberton, *A View of Sir Isaac Newton's Philosophy* (London: S. Palmer, 1728), p. 28.
Our quotation was taken from p. "a" of the Preface to this short work. There is an interesting discussion on the next page of the way Newton recognized the so called D^2 law relating the inverse relationship between the amount of gravitational force and the square of the distance between the bodies involved.
7. François Marie Arouet Voltaire, *Elements de la Philosophie de Newton, Oeuvres Complètes de Voltaire. Tome Cinquième, Imprimerie et Fonderie d'Everat* (Paris, 1836).
The original edition was printed in Amsterdam in 1738. Our notes were translated by George C. S. Adams and Dan Welch from p. 727, part 3, chap. 3. Madame Conduit, Newton's niece, gave Voltaire the information; he gave no details of the conversations he must have had

with her. Physics teachers will find this paper a most stimulating reflection on both Newton and the birth of science.

8. Kevin Ballard, Premedical student, Wofford College, Spartanburg, S. C. Personal communication, 1979.
9. Earle Scott, Premedical student, Wofford College, Spartanburg, S.C. Personal communication, 1979.
10. Daniel Welch, Assistant Professor of Physics, Wofford College, Spartanburg, S.C. Personal communication, 1979.
11. Audrey Bergerson Eddy, Consultant, South Carolina Department of Education, Greenville, S.C. Personal communication, 1979.
12. John W. Harrington, Professor of Geology, Wofford College, Spartanburg, S.C., who never became an experimental scientist. Personal communication.
13. James E. Seegars, Jr., Professor of Psychology, Wofford College, Spartanburg, S.C. Personal communication, 1979.
14. Linus Pauling, Linus Pauling Institute of Science and Medicine, Menlo Park, Calif. Personal communication, 1979.
15. Howard A. Meyerhoff, Consulting geologist, Tulsa, Okla. Personal communication, 1979.
16. Henry S. Johnson, Jr., Consulting geologist, Charleston, S.C. Personal communication, 1979.
17. Rudyard Kipling, *Departmental Ditties and Ballads and Barrack Room Ballads* (Garden City, N.Y.: Doubleday, 1914).
    The Lost Legion is a surprisingly deep poem, for it establishes the catlike quality necessary for trailbreakers. This same theme is found in many other Kipling poems. They ring true, in part, at least because Kipling was a product of the frontier and knew firsthand the strength of the men and women who lived there. This poet is often dismissed as someone who can speak only of the "white man's burden." Unfortunately, that phrase from the title of one of his frontier poems has tagged his literary image and he is no longer read in depth.

SUBJECT GUIDE

Page number in boldface indicates that an illustration appears on that page.

ABCDEFGHIJ-BP-8210